Electrical Estimating Methods

Electrical Estimating Methods

Fourth Edition

WAYNE J. DEL PICO, CPE

RSMeans
FROM THE GORDIAN GROUP

WILEY

Cover design: Wiley
Electrical box Image: © knowlesgallery/Thinkstock | Electrical blueprint image: © OLJensa/
Thinkstock; Circuit diagram Image: © Monkey Business Images/
Thinkstock | Worker Image: © Ingram Publishing/Thinkstock

This book is printed on acid-free paper. ∞

For general information about our other products and services, please contact our Customer Care
Department within the United States at (800) 762-2974, outside the United States at (317) 572-3993
or fax (317) 572-4002.

Wiley publishes in a variety of print and electronic formats and by print-on-demand. Some material
included with standard print versions of this book may not be included in e-books or in print-on-
demand. If this book refers to media such as a CD or DVD that is not included in the version you
purchased, you may download this material at http://booksupport.wiley.com. For more information
about Wiley products, visit www.wiley.com.

Library of Congress Cataloging-in-Publication Data:

Del Pico, Wayne J.
    [Means electrical estimating methods]
    Electrical estimating methods / Wayne J. Del Pico. — Fourth edition.
      pages cm
    Original ed. published under title: Means electrical estimating methods. c1995.
    Includes index.
    ISBN 978-1-118-76698-9 (paperback); ISBN 978-1-118-76684-2 (ebk.); ISBN 978-1-118-76696-5
(ebk.); ISBN 978-1-118-96323-4 (ebk.)
    1. Electrical engineering—Estimates. I. R.S. Means Company. II. Title.
  TK435.M428 2014
  621.319'24029—dc23

                                                              2014018376

Printed in the United States of America

SKY10068383_022724

Dedicated to the memory of

Sid Numerof

1929–2013

Good friend, valued coworker, and dedicated family man

Contents

About the Author

Wayne J. Del Pico is president of W. J. Del Pico, Inc., where he provides construction management and litigation support services for construction related matters. He has more than 35 years of experience in construction project management and estimating and has been involved in projects throughout most of the United States. His professional experience includes private commercial construction, public construction, retail construction, and residential land development and construction.

Mr. Del Pico holds a degree in civil engineering from Northeastern University in Boston, where he taught construction-related curriculum in Cost Estimating, Project Management, and Project Scheduling from 1992 until 2006. He is also a member of the adjunct faculty at Wentworth Institute of Technology in Boston, where he presently teaches programs in Construction Cost Analysis, Estimating, Project Control, and Construction Scheduling.

Mr. Del Pico is a seminar presenter for the RSMeans Company, where he provides instruction on topics from estimating to scheduling. He is the author of *Plan Reading and Material Takeoff* (1994), *Estimating Building Costs (2004)* and its second edition in 2012, and is a co-author of *The Practice of Cost Segregation Analysis (2005)*. His most recent book, *Project Control: Integrating Cost and Schedule in Construction,* was published by Wiley in September 2013.

His construction experience and knowledge of the industry has qualified him to be the past president of the Builders Association of Greater Boston (2010). He is also a practicing Neutral for the American Arbitration Association, where he hears construction-related arbitration cases.

To learn more about the author, visit www.wjdelpico.com.

1 | THE ESTIMATING PROCESS

1 | Components of an Estimate

One of the most difficult tasks in estimating any project is to capture all of the costs involved in the project. Construction has many variables, and it is these variables that can have an impact of the way the estimator "sees" the work and ultimately its costs. The means and methods selected, or the plan to execute the work, will impact price significantly. Another important variable is the bid documents; comprehensive, fully developed designs offer a better chance for the estimator to reach an accurate price. It is the goal of the estimator to arrive at the most accurate price for the cost of the work under a specific set of circumstances and conditions.

While different estimators may see a project differently and thereby arrive at a different price for the work, all estimates share some basic components. Every cost estimate requires three basic components. The first is the establishment of standard *units of measure*. The second component of an estimate is the determination of the *quantity* of units for each component, which is an actual measurement process: how many linear feet of wire, how many device boxes, and so on. The third component, and perhaps the most difficult to obtain, is the determination of a reasonable *cost* for each unit.

The first element, the designation of measurement units, is the step that determines and defines the level of detail, and thus the degree of accuracy, of a cost estimate. In electrical construction, such units could be as all-encompassing as the number of watts per square foot of floor area or as detailed as a linear foot of wire. Depending on the estimator's intended use, the designation of the unit of measure may describe a complete system, or it may be a single task within the entire scope of the project. The selection of the unit of measure also determines the time required to do the estimate.

The second component of every estimate, the determination of quantity, is more than simply counting units. In construction, this process is called the *quantity take-off* or *quantity survey*. It is an integral part of the estimating process that requires an intimate understanding of the work being estimated and a commitment to accuracy. To perform this function successfully, the estimator should have a working knowledge of the materials, methods, and codes used in electrical construction. An understanding of the technical specifications that were the basis of the design is also particularly important. This knowledge helps to ensure that each quantity is correctly tabulated and that essential items are not forgotten or omitted. The estimator with a thorough knowledge of construction is also more likely to account for all requirements in the estimate.

A clear understanding of the *scope*, or limits of the work, what is included and what is not, is also critical for a defining the estimate.

Not all of the tasks in an estimate involve materials; some are labor-only tasks. Testing is an example of a labor-only item. Some can be just material items, for example, a work box and conduit that is set in a masonry wall by the bricklayer. Experience is, therefore, invaluable to ensure a complete estimate.

The third component is the determination of a reasonable cost for each unit referred to as *pricing*. This aspect of the estimate is significantly responsible for variations in estimating. Rarely do two estimators arrive at exactly the same material cost for a project. Even if material costs for an installation are the same for competing contractors, the labor costs for installing that material can vary considerably, as a result of varying productivity and pay scales in different areas. The use of specialized equipment can decrease installation time and, therefore, cost. Finally, material prices fluctuate within the market. These cost differences occur from city to city and even from supplier to supplier in the same town. It is the experienced and well-prepared estimator who can keep track of these variations and fluctuations and use them to his or her best advantage when preparing accurate estimates.

This third phase of estimating, the determination of costs, can be defined in three different ways by the estimator. With one approach, the estimator uses a unit cost that includes all the elements (i.e., material, installation, overhead, and profit) in one number expressed in dollars per unit of work. A variation of this approach is to use a unit cost that includes total material and installation as a single amount, adding a percent markup for overhead and profit in the estimate summary.

A second method is to use individual unit costs for material and for installation. Costs are calculated separately for each category without markups. These are called *bare costs*. Different profit and overhead markups are applied to each item before the material and installation prices are totaled. The result is called the *billing* rate or price.

A third method of pricing uses unit costs for materials, with labor-hours as the measure of labor. Again, these figures are totaled separately; one represents the value of materials expressed in dollars, and the other shows the total labor-hours for installation. The average cost per hour of trade labor is determined by allowing for the expected ratios of foremen, journeymen, and apprentices. This is sometimes called a *composite labor rate*. This rate is multiplied by the total labor-hours to get the total bare cost of installation. Different overhead and profit markups can then be applied to each, material and labor, and the results added to get the total billing rate.

Whichever methodology is selected, it is important to remember that it should remain consistent through the entire estimate to avoid errors, omissions, or duplications. The estimator must, therefore, exercise care to utilize these methods correctly and consistently for the format of each particular estimate.

As a point of clarification, the word *unit* is used in many ways, as can be seen in the preceding definitions. Keeping the concepts of units clearly defined is vital to achieving an accurate, professional estimate. For the purposes of this book, the following references to different types of units are used:

- **Unit of measure.** The standard by which the quantities are counted, such as *linear feet* of conduit, or *number* of boxes. There are industry-accepted standards of units for almost all work.
- **Cost units.** The total dollar price per each installed unit of measure, including the costs of material and installation. This figure may be a bare cost or may include overhead and profit.
- **Material unit cost.** The cost to purchase each unit of measure. This cost represents material dollars only—with no overhead and profit.
- **Installation unit cost.** The cost for installing each unit of measure. This cost includes labor dollars only—with no overhead and profit.
- **Labor unit.** The labor-hours required to install a unit of measure. (*Note:* Labor units multiplied by the labor rate per hour equals the installation unit cost in dollars.)

A final thought on cost: It is important to note that the word *cost* is defined by its frame of reference. For the general contractor; the electrical contractor's entire price is a cost. When the work is complete, the general contractor will pay the entire contract amount to the electrical contractor and record it as a cost to the project. For the electrical contractor, cost is defined as all amounts in the estimate, with the *exception* of the profit. The electrical contractor will records costs as material, labor, and equipment paid to others, while the profit made is the only item not classified as a cost.

2 | Types of Estimates

Estimators use four basic types of estimates. These types may be referred to by different names and may not be recognized by all as definitive. Most estimators, however, will agree that each type has its place in the construction estimating process. The four types of estimates are as follows:

- **Order of magnitude estimate.** The order of magnitude estimate could be loosely described as an educated guess. It can be completed quickly. Accuracy will vary between 20% and 25%.
- **Square foot estimate.** This type is most often useful when only the proposed size and use of a planned building is known. This method can be completed within an hour or two. Accuracy can be plus or minus 15%.
- **Assemblies estimate.** A systems estimate is best used as a budgetary tool in the planning stages of a project when some parameters have been decided. This type of estimate could require as much as one day to complete. Accuracy is expected to be plus or minus 10%.
- **Unit price estimate.** Working drawings and full specifications are required to complete a unit price estimate. It is the most accurate of the four types but is also the most time consuming. Used primarily for bidding purposes, the accuracy of a unit price estimate can be plus or minus 5%.

As an estimator *and* his or her company gain repetitive experience on similar or identical projects, the accuracy of all four types of estimates will improve dramatically. In fact, given enough experience and the historical data backup, *square foot* estimates can be extremely accurate for certain types of work. However, most prudent contractors would never sign a contract based on a square foot price for the electrical scope of the work without some wiggle room. Unit price estimates are still the method of choice for competitive bidding leading to contract.

ORDER OF MAGNITUDE ESTIMATES

The order of magnitude estimate, also called a *rough order of magnitude (ROM) estimate*, can be completed with a minimum amount of information and a small expenditure of time. The units of measure, described in Chapter 1, "Components of an Estimate," can be very general for this type of estimate and require little definition. The units of measure are frequently units not typical to the construction industry and are used for cost-benefit analysis and very early decision making. For example, the cost of electrical work for a residential apartment building can be provided in a cost per apartment.

This type of ROM estimate can be made after a few minutes of analysis, drawing on experience and historical data from similar past projects. While this ROM might be appropriate for initial decision making, it does not take into account the uniqueness of individual projects. Experienced electrical contractors with historical data from similar projects can distill the total project cost into units of measure that are at their most basic. For example, the total electrical cost for an apartment complex could be provided in terms of the number of apartments in the complex. For parties with no historical cost data from which to draw, there are sources of published cost data that can provide data that can be the basis of a ROM estimate.

Table 2.1 and 2.2, from *Means Electrical Cost Data*, is a source of data that can be used in generating early ROM estimates. As previously stated, this cost data is in a unit of measure that is representative of the type and use of the project. As an

Table 2.1   Order of Magnitude Data (Lines 9000 and 9500)

50 17 | Square Foot Costs

		50 17 00 | SF Costs	UNIT	UNIT COSTS			% OF TOTAL			
				1/4	MEDIAN	3/4	1/4	MEDIAN	3/4	
01	0010	**APARTMENTS Low-Rise (1 to 3 story)**	SF	73	92.50	123				01
	0020	Total project cost	CF	6.55	8.70	10.75				
	0100	Site work	SF	5.35	8.55	15	6.05%	10.55%	13.95%	
	0500	Masonry		1.44	3.55	5.80	1.54%	3.92%	6.50%	
	1500	Finishes		7.75	10.65	13.15	9.05%	10.75%	12.85%	
	1800	Equipment		2.40	3.63	5.40	2.71%	3.99%	5.95%	
	2720	Plumbing		5.70	7.30	9.30	6.65%	8.95%	10.05%	
	2770	Heating, ventilating, air conditioning		3.63	4.47	6.55	4.20%	5.60%	7.60%	
	2900	Electrical		4.25	5.65	7.65	5.20%	6.65%	8.35%	
	3100	Total: Mechanical & Electrical		15.10	19.60	24	16.05%	18.20%	23%	
	9000	Per apartment unit, total cost	Apt.	68,000	104,000	153,500				
	9500	Total: Mechanical & Electrical	"	12,900	20,300	26,500				

example, refer to the bottom of the category titled *APARTMENTS Low-Rise (1 to 3 Story)*. The proposed use and magnitude of the planned structure—such as the desired number of apartments in an apartment complex—may be the only parameters known at the time the ROM Estimate is done. The data given in Table 2.2 does not require that details of the proposed project be known to determine rough costs; the only required information is the intended use and capacity of the building. The lack of accuracy can be subsidized with the addition of a contingency of 20% to 25%.

SQUARE FOOT ESTIMATES

Another type of estimate requires more definition to the project. In addition to the building's use or type, the definition is provided in the form of its size in gross square area of the building. This type of estimate is called the *square foot estimate*. The use of square foot estimates is most appropriate after the conceptual design has been started and maybe only a floor plan and elevation exist, although these types of estimates can be applied in the absence of any plans. This allows early cost estimates to be generated and budgetary parameters to be set.

For the electrical contractor with the historical data and experience, he or she can translate total project costs into dollars per gross square foot of building. The best source of square foot costs is the estimator's own cost records for similar projects, adjusted to the parameters of the project at hand. Once again, this is a preliminary estimate and not meant to be the cost basis of a contract.

Table 2.2   Square Foot Cost Data

50 17 | Square Foot Costs

	50 17 00	SF Costs		UNIT	UNIT COSTS			% OF TOTAL			
					1/4	MEDIAN	3/4	1/4	MEDIAN	3/4	
01	0010	**APARTMENTS Low-Rise (1 to 3 story)**		SF	73	92.50	123				01
	0020	Total project cost		CF	6.55	8.70	10.75				
	0100	Site work		SF	5.35	8.55	15	6.05%	10.55%	13.95%	
	0500	Masonry			1.44	3.55	5.80	1.54%	3.92%	6.50%	
	1500	Finishes			7.75	10.65	13.15	9.05%	10.75%	12.85%	
	1800	Equipment			2.40	3.63	5.40	2.71%	3.99%	5.95%	
	2720	Plumbing			5.70	7.30	9.30	6.65%	8.95%	10.05%	
	2770	Heating, ventilating, air conditioning			3.63	4.47	6.55	4.20%	5.60%	7.60%	
	2900	Electrical			4.25	5.65	7.65	5.20%	6.65%	8.35%	
	3100	Total: Mechanical & Electrical			15.10	19.60	24	16.05%	18.20%	23%	
	9000	Per apartment unit, total cost		Apt.	68,000	104,000	153,500				
	9500	Total: Mechanical & Electrical		"	12,900	20,300	26,500				

Source: Reprinted with permission from Reed Construction Data from *RSMeans Electrical Cost Data 2014.*

For parties with no historical cost data of their own, published cost data can serve as the basis of the estimate. Referencing the *APARTMENT Low-Rise (1 to 3 Story)* category in Table 2.2, one will note that costs are presented first as *total project costs* by square foot. These costs are broken down into major categories of work in terms of costs per square foot, and then into the relationship of each category to the project as a whole, in percentages. This breakdown enables the estimator to adjust certain categories of work according to the unique requirements of the proposed project. The costs on this and other pages of *Means Electrical Cost Data* are representative of actual project costs contained within the RSMeans database. These costs include the contractor's overhead and profit but do not include architectural fees, carrying costs, or other soft costs. The 1/4 column shows the value at which 25% of the projects had lower costs and 75% had higher. The 3/4 column value denotes that 75% of the projects had lower costs and 25% had higher. The median column value shows that 50% of the projects had lower costs and 50% had higher.

While helpful for preparing preliminary budgets, square foot estimates can also be useful as checks against other, more detailed estimates—a "big-picture" check and balance. While more time is required than with ROM estimates, a greater accuracy is achieved because of a more specific definition of the project.

ASSEMBLIES ESTIMATES

The next level on the evolutionary scale of estimating is the *assemblies estimate*. This method categorizes the estimate into major systems of the structure. The assemblies estimate provides the distinct advantage of enabling alternate construction techniques to be readily compared for budgetary purposes. Rapidly changing construction costs in recent years have made budgeting and cost-effectiveness studies increasingly important in the early stages of building projects. Unit price estimates, because of the time and detailed information required, are not well suited as budgetary or planning tools. A faster and more cost-effective method with the needed flexibility was created for the planning phase of a building project. It is the assemblies estimate.

The assemblies method is a logical, sequential approach that reflects how a building is constructed. The estimate is organized into seven groups based on the major components that can be found in a project. This organizational structure is called *UNIFORMAT II* and is outlined as follows:

Assemblies Major Groups

A—Substructure

B—Shell

C—Interiors

D—Services

E—Equipment & Furnishings

F—Special Construction

G—Building Sitework

Each major group is further broken down into systems. *Division D, Services*, which covers electrical construction, is composed of the following groups of systems:

D5010—Electrical Service & Distribution

D5020—Lighting & Branch Wiring

D5030—Communications & Security

D5090—Other Electrical System

Each system incorporates several different components into an assembly that is commonly used in construction.

A great advantage of the assemblies estimate is that the estimator is able to substitute one system or assembly for another during design development and can quickly determine the relative cost differential. This allows the decision makers for the project to determine the benefit of one system over another. The owner can use the assemblies estimate as guidance in keeping the project on a budgetary tract.

The assemblies method does not require the degree of final design details needed for a unit price estimate, but estimators who use this approach must have a solid background knowledge of construction materials and methods, code requirements, design options, and budget considerations.

The assemblies estimate should not be used as a substitute for the unit price estimate. While the assemblies approach can be an invaluable tool in the planning stages of a project, it should be supported by unit price estimating whenever greater accuracy is required.

UNIT PRICE ESTIMATES

At the top of the evolutionary scale of estimating is the *unit price estimate*. This method is the most accurate and detailed of the four estimate types and therefore takes the most time to complete. It is a decomposition of the design into incremental units called *tasks* or *activities*. It requires detailed working drawings and specifications as a basis of the estimate. All decisions regarding the project's materials and systems must have been made to complete this type of estimate. The lack of variables provides the basis for a more accurate estimate than with any of the previous methods.

Working drawings and specifications are used to determine the quantities of materials, equipment, and labor. Current and accurate unit costs for these items are also necessary. The most accurate cost basis is always historical data collected and analyzed from previous projects of a similar type. Wherever possible, estimators should use prices based on experience or developed from actual, similar projects.

In the absence of a historical basis, costs can come from published data, such as *Means Electrical Cost Data*.

Because of the detail involved and the need for accuracy, completion of a unit price estimate entails a great deal of time and expense. For this reason, unit price estimating is best suited for construction bidding. It can also be an effective method for determining certain detailed costs in a conceptual budget or during design development.

The organization of the unit price estimate follows an industry recognized format called *CSI MASTERFORMAT 2010*™. CSI MASTERFORMAT 2010 was developed by *Construction Specifications Institute, Inc.* or *CSI* and is an expansion of the original CSI MASTERFORMAT that has been used and accepted for years. CSI grouped similar types of work into *divisions*. In an effort to accommodate the changes in technology, the original version has been expanded to 50 divisions. The outline of CSI MASTERFORMAT 2010 is as follows:

MasterFormat Divisions

Division 00—Procurement and Contracting Requirements

Division 1—General Requirements

Division 2—Existing Conditions

Division 3—Concrete

Division 4—Masonry

Division 5—Metals

Division 6—Wood & Plastics

Division 7—Thermal & Moisture Protection

Division 8—Openings

Division 9—Finishes

Division 10—Specialties

Division 11—Equipment

Division 12—Furnishings

Division 13—Special Construction

Division 14—Conveying Systems

Divisions 15–20—*Reserve divisions for future expansions*

Division 21—Fire Suppression

Division 22—Plumbing

Division 23—HVAC

Division 24—*Reserve division for future expansion*

Division 25—Integrated Automation

Division 26—Electrical

Division 27—Communications

Division 28—Electronic Safety and Security

Divisions 29 and 30—*Reserve divisions for future expansions*

Division 31—Earthwork

Division 32—Exterior Improvements

Division 33—Utilities

Division 34—Transportation

Division 35—Waterway and Marine Construction

Divisions 36–39—*Reserve divisions for future expansions*

Division 40—Process Integration

Division 41—Material Process and Handling Equipment

Division 42—Process Heating, Cooling & Drying Equipment

Division 43—Process Gas & Liquid Handling, Purification and Storage Equipment

Division 44—Pollution Control Equipment

Division 45—Industry-Specific Manufacturing Equipment

Division 46—Water and Wastewater Equipment

Division 47—*Reserve division for future expansion*

Division 48—Electrical Power Generation

Division 49—*Reserve division for future expansion*

CSI MASTERFORMAT 2010™ is categorized into five subgroups:

General Requirements Subgroup—Division 1

Facilities Construction Subgroup—Divisions 2–19

Facilities Services Subgroup—Divisions 20–29

Site and Infrastructure Subgroup—Divisions 30–39

Process Equipment Subgroup—Divisions 40–49

Each of the divisions is further divided into subsections, as a way of refining the categorization of work. For example, consider *Division 26—Electrical:*

26 01—Operation and Maintenance of Electrical Systems

26 05—Common Work Results for Electrical

26 06—Schedules for Electrical

26 08—Commissioning of Electrical Systems

26 09—Instrumentation and Controls for Electrical Systems.

26 10—Medium-Voltage Electrical Distribution

26 20—Low-Voltage Electrical Transmission

26 30—Facility Electrical Power Generating and Storing Equipment

26 40—Electrical and Cathodic Protection

26 50—Lighting

The CSI method of organizing the various components provides a standard of uniformity that is widely used by construction industry professionals: contractors, material suppliers, engineers, and architects.

A sample unit price page from *Means Electrical Cost Data* is shown in Table 2.3. This page lists various types of interior light fixtures each with a unit price for each type of fixture. In the absence of historical data, publish cost data can provide a database of information useful in developing a Unit Price Estimate. The type of work to be performed is described in detail: typical crew make-ups, unit labor-hours, units of measure, and separate costs for material and installation. Total costs are extended to include the installing contractor's overhead and profit.

As a subset to the unit price estimate, another type of estimate warrants an introduction. The *scheduling estimate* involves the application of labor allocations. A complete unit price estimate is a prerequisite for the preparation of a scheduling estimate. While the discussion of scheduling estimates goes beyond the scope of this book, an introduction to the practice will be addressed in Chapter 9, "Prebid Scheduling."

Table 2.3   Unit Price Costs Data

26 51 Interior Lighting

26 51 13 - Interior Lighting Fixtures, Lamps, and Ballasts

26 51 13.50 Interior Lighting Fixtures		Crew	Daily Output	Labor-Hours	Unit	2014 Bare Costs				Total Incl O&P
						Material	Labor	Equipment	Total	
5500	12", four 60 watt lamps	1 Elec	6.70	1.194	EA	69.50	63.50		133	172
5510	Pendant, round, 100 watt		8	1		111	53.50		164.50	202
5520	150 watt		8	1		121	53.50		174.50	213
5530	300 watt		6.70	1.194		169	63.50		232.50	282
5540	500 watt		5.50	1.455		320	77.50		397.50	465
5550	Square, 100 watt		6.70	1.194		149	63.50		212.50	260
5560	150 watt		6.70	1.194		156	63.50		219.50	267
5570	300 watt		5.70	1.404		227	75		302	360
5580	500 watt		5	1.600		310	85.50		395.50	470
5600	Wall, round, 100 watt		8	1		64.50	53.50		118	151
5620	300 watt		8	1		113	53.50		166.50	204
5630	500 watt		6.70	1.194		375	63.50		438.50	505
5640	Square, 100 watt		8	1		102	53.50		155.50	192
5650	150 watt		8	1		104	53.50		157.50	194
5660	300 watt		7	1.143		164	61		225	272
5670	500 watt		6	1.333		288	71		359	420

Source: Reprinted with permission from Reed Construction Data from *RSMeans Electrical Cost Data 2014*.

3 | Before Starting the Estimate

In the competitive world of construction contracting an invitation to bid a project can mean the prospect of weeks of hard work with only a chance of bidding success. It can also mean the opportunity to obtain a contract for a successful and financially beneficial project. It is not uncommon for the contractor's bidding schedule to include multiple bids lost for every successful one that goes to contract. The successful comprehensive bid depends heavily on estimating accuracy and thus the preparation, organization, and work that goes into estimating a project. After all, nobody really wants the project that was won due to errors or omissions in the estimate.

OBTAINING BID DOCUMENTS

The first step in starting the estimating process is to obtain copies of the bid documents; plans, addenda, and the project manual in its *entirety*. Long past are the days when the general contractor (GC) provides only the trade's drawings and specifications and requests a comprehensive bid covering all of the work. This is especially true for the electrical scope of work, which has an almost continual interface on the project. It is essential that the bidding electrical estimator obtain a *full set* of the bid documents.

For a major portion of commercial projects, bid documents are hosted on an *ftp* or *File Transfer Protocol* site, where they can be viewed and downloaded for estimating purposes. The electrical contractor can then decide which drawings, if any, to print for estimating. This avoids the clerical error of not printing what is needed by a nonestimator. The ftp site allows the bidding contractor to review and retain an entire set of the project bid documents. The helps prevent major tasks in the scope of work from being omitted as a result of not having the full set.

TO BID OR NOT TO BID

Not every project we are asked to bid is a fit for the company. Numerous considerations must be reviewed before providing a commitment of resources and funds to bid a project. A thorough and comprehensive review of the bid documents is essential. For the seasoned professional, this review starts with a careful study of the proposed contract terms and language that is contained in the project manual. If the contractor cannot accept the terms and conditions offered in the contract language for things such as payment or dispute resolution, it might be best to decline to bid the project. Most owners are reluctant to change the contract terms and conditions unless an overwhelming majority of bidders decline to bid the project. In today's litigation-prone construction industry, many owners and design professionals produce contracts with exculpatory language that is so biased that many contractors are required to accept a disproportional share of the risk with little or no control. As a result, many contractors, GC and subcontractor alike, are not willing to accept the risk. A special note should be added here: the decision to bid may be done at a senior management level and not within the purview of the estimator. This is dependent on the management structure of the company.

A similar concern holds true for projects with defective documents. More than a few of the project documents put out for bids are underdeveloped and are ripe for a legal battle. Restrictive language is substituted for a comprehensive design that is expected to transfer the burden to the prime and subcontractors. The estimator has a duty to advise management of the quality of the bid documents, especially if there is potential to cause financial harm to the company.

The second step in deciding-to-bid process is to check the bid date. The bidder must evaluate whether there is sufficient time in which to prepare a proper bid. Sufficient time, not just for the estimator, but for the vendors, sub-subcontractors and other parties that will provide contemporaneous quotes. Hastily prepared estimates are candidate for errors or omissions. Other consideration include:

- Current workload and available resources
- Expertise of personnel in the type of electrical work
- Location of work within our normal area
- Monetary size of work within our means to capitalize
- Sufficient workforce to execute within the schedule
- Schedule available reasonable and possible
- Prior experience with the design professionals, GC, or owner
- Bonding and insurance requirements

DOCUMENT STUDY

Once the decision to bid has been made, the next step is a thorough *study* of the bid documents. A study is substantially more than a review and requires a drawing-by-drawing analysis of the plans and a page-by-page reading of the specifications. This is a time-consuming but essential step in becoming familiar with the project. It should be done, if possible at a single sitting, in a quiet environment allowing for concentration. The estimator should have a way to make notes during the study. As the study progresses, many of the notes or questions are answered. It is the final ones that may be the basis of a *Request for Information* or *RFI*.

This initial study has several distinct benefits in addition to the obvious familiarization process, one of which is that the study forces a detailed thinking of the work. This is the point in which the estimator considers the best "means and methods" to accomplish the work. He or she also evaluates whether the work can be done as designed. Another benefit is that the estimator evaluates which other parties need to be involved; supply houses, subcontractors, testing companies, equipment suppliers, and even which supervisor would be best for the project. This is a valuable step in estimating that gets done early in the process.

For estimators that use paper plans, notes, and references can be written directly on the sheets. Bid documents, especially the set with the estimator's notes must be retained even if the bid is not successful. Individual companies will have varying policies regarding the length of time for storage of the physical plans and paper files. For estimators that employ a software system for takeoff, notes can be added electronically and placed directly within the design in much the same manner as one would do with paper plans. In this application the entire set of bid documents along with the takeoff, can be stored electronically for an indefinite period since physical space is not a concern. The same rules apply to projects that are bid but never awarded to any company. Occasionally, these projects come back to life, and it is beneficial to have prior documents for comparison.

Once the study has been completed, the site visit can be scheduled. For projects with a prescheduled or mandatory site visit, it is still best practice to do the document study in advance. This has the benefit of providing the estimator a frame of reference for what he or she will see on the site visit. In addition, an estimator that is thoroughly familiar with the project asks questions that are beyond what can be found in the documents and often conveys a higher level of preparedness and expertise to those conducting the site visit. In some cases, it is the basis of the initial impression that is so important. During a site visit, the estimator should take notes, and possibly photographs, of all situations pertinent to the construction and thus to the project estimate. If unusual site conditions exist, or if questions arise during the takeoff, a second site visit is recommended.

The next step in the preparing-to-bid is adding the *addenda* or changes to the documents. Addenda are often issued *after* the bid documents are released. Addendum can be as simple a clarification of a detail on the plans or a completely new sheet. It can include changes or additions to the technical specifications, or additional sections to the project manual. Regardless of the specific change, the plans and the specifications must be brought up to date or *addendumized*. This process can include the physical or electronic cutting and pasting of sketches or notes onto the plans or in the specifications. This allows the estimator to have the most up-to-date documents at his or her fingertips when preparing the bid. It reduces the search time for information and avoids the error caused by an omission of information contained within the addenda.

ESTIMATING TEAM MEETING

For the estimator lucky enough to have assistance, a meeting should be conducted with other members of the estimating team that will help in the preparation of the estimate. Project needs for vendors, subcontractors, suppliers, and the like should be addressed along with any unique conditions of the project or requirements under the contract. Everyone should be made aware of the bid date and time. A brief scope review is also beneficial for the other team members so all know what is included and what is not. The related work of other parties is also discussed to establish limits of work and coordination. A list of the subs, vendors, and suppliers that will be used for pricing are discussed and invitations assigned to be sent.

The bid form, if included, is reviewed for signature requirements or update documents to be included with the bid or any other specific compliances that can be done in advance of the bid date.

Bonding requirements, both for the bid bond and the performance and payment bonds, should be addressed early with the surety. Many a surety has declined to provide a bond after the principal has invested time in the estimating process.

The due date should be marked on a calendar and a schedule. Completion of the takeoff should be made as soon as possible, not just prior to the bid deadline.

Insurance policies and limits should be reviewed to ensure that the company is covered. If additional coverage must be added if awarded the work, early acknowledgment gives the insurance agent adequate time to get competitive bids for the coverage so that the cost can be included in the estimate.

It is important that all team members become familiar with the project as soon as possible. Responsibilities are assigned and milestones for specific phases of the estimate are established.

As guidance, some contractors use a checklist system for these meetings to streamline them while ensuring all points are adequately touched upon. Division 00—Procurement and Contracting Requirements section of the project manual is often a good resource for developing a checklist.

Final procedures include a review of the Supplemental General Conditions of the Contract. These modify the General Conditions of the Contract for Construction, the most widely recognized of which is the AIA A201. Supplemental Conditions can have a tremendous impact of the terms and conditions of a contract and as such can affect the way a project is estimated and ultimately bid.

Also included in the project manual is information regarding completion dates, payment schedules (e.g., retainage), submittal requirements, allowances, alternates, and other important project requirements. Each of these conditions can have a significant impact on the ultimate cost of the project. They must be read and understood prior to performing the estimate.

4 The Quantity Takeoff

The quantity survey or *takeoff* is the starting point for any estimate. The takeoff is the decomposition of the work shown on the plans and specifications into units of measure that can be priced. The amount of detail shown on electrical drawings is perhaps the least of any for the major trades. Much of the work is shown with symbols and graphics that illustrate the design. As a result, the electrical estimator must not only count the items shown on the drawings but also envision the completed work, including fittings, hangers, fasteners, devices, and cover plates. A precise and thorough quantity takeoff is the basis for a sound estimate. Errors or inaccuracies in this portion of the estimate are compounded during the pricing phase, regardless of how reliable the unit prices are.

The takeoff should be organized so that the information gathered can be used in other parts of the construction process. For example, a schedule for performing the work can be created from the estimate, specifically crew composition, labor-hours, and task definition. It should also be noted that if the bid is successful and the work is awarded, the estimate will serve as the basis for the performance measurement baseline for project control. Even material procurement and equipment rental terms will have their genesis in the takeoff.

In this chapter, we will explore the most common practices of taking off quantities for an electrical construction project and offer suggestions for developing routine procedures that will help ensure accuracy.

BEGINNING THE QUANTITY TAKEOFF

The quantity takeoff breaks the project down into its elemental parts, called *tasks* or *activities*. (For the purpose of estimating, these terms are interchangeable.) The quantity takeoff is not a list of materials required, but a series of tasks that are required to complete the contractual obligations. Tasks are actual units of work to be performed, such as: *Furnish and install 2" PVC conduit,* or *Install 2x4 light fixture at hallways,* or *Load Test Panel P2.* They are quantified by details and dimensions provided on the drawings. Tasks consume resources, cost money, and take time out of the schedule. Tasks are also identified in terms of the quality of materials and labor required. This level is established in the specifications. Establishing the quantity and quality of a task is an essential part of accurate estimating. Without them, the estimate is nothing more than a guess, which could cause conflict if different methods or materials are required by the engineer or owner when the construction work is performed.

Tasks or activities are made up of various components, including material, labor, tools and/or equipment, subcontractors, and occasionally non-production-related expenses, such as electrical permits, bond costs, and direct overhead costs. Not every task includes all of these components. Determining which components apply is a combination of what the documents call for, along with judgment and experience.

Traditionally, quantities are taken off from the drawings in roughly the same sequence as they are erected or installed. The takeoff can also be arranged by a major system or phase of the work. Many estimates contain a combination of these two.

Ideally, quantities should be taken off by one person if the project is not too large and if time allows. This same individual should also do the pricing. This reduces errors that occur from misinterpretation of another's work. For larger projects, the takeoff responsibilities are often shared and the work assigned to two or more quantity surveyors. In this case, a project leader should be assigned to coordinate and assemble the estimate.

It should be noted that the previously mentioned general points apply regardless of whether the estimator is doing the takeoff "by hand" or with computer-based software.

To effectively, efficiently, and accurately perform the takeoff, certain guidelines should be followed.

RULES TO FOLLOW FOR ACCURATE TAKEOFFS

The following takeoff rules are based on common sense practices that will help prevent, or at least minimize, errors. They can help make the takeoff better organized, more efficient, and more accurate. Again, these rules apply regardless of whether the estimator is doing the takeoff "by hand" or with computer-based software.

Rule 1: Write Clear Task Descriptions

Descriptions should be clear and legible and should indicate the work needed and the part of the structure involved—or the location on the drawing where the quantity originates. When taking off quantities, descriptions should be written according to the guidelines of the individual who will apply the unit prices.

Rule 2: Use Industry-Accepted Units

A takeoff is not a list of materials for use in placing an order, but rather a descriptive list of activities with quantities derived from the dimensions on the documents.

These quantities must then be extended into industry-accepted units for pricing. For example, conduit is estimated in linear feet (LF) or 100 linear feet (CLF) because that is how it is sold. LF and CLF are also the accepted industry standard units.

Rule 3: Follow a Logical Order

The takeoff should be logical and organized. The best approach is to proceed in roughly the same order as one would perform the work. This allows you to visualize the process while performing the takeoff. The logical thought process is to consider "what is the next step?" and organize the takeoff according to the CSI MASTERFORMAT classification system. (See Chapter 2, "Types of Estimates", for more on the MASTERFORMAT.)

Rule 4: Review Scales, Notes, Abbreviations, Symbols, and Definitions

Review drawings and details carefully for notes and scale. Scale can change from drawing to drawing, and a general rule of thumb is that as the detail becomes smaller in focus, the scale becomes larger.

It is necessary to become familiar with symbols and abbreviations that typically appear on the drawings. Frequently, drawings contain legends that define material and graphic symbols. Abbreviations, such as *NTS* (not to scale) and *TYP* (typical), may be used throughout the entire set of plans. Some words in the construction industry have unique meanings, such as *provide*, which is defined as *"to furnish and install."* Carefully review any specification sections that include references or definitions, as these can have a large impact on the estimate.

Rule 5: Verify Dimensions

Wherever possible, use the dimensions exactly as they appear on the drawings. Add intermediate dimensions to arrive at total dimensions. Scale dimensions should be used only as a last resort. Develop the habit of checking printed dimensions against scaled dimensions. Discrepancies should be brought to the attention of the design team or the owner.

Always express dimensions in the same order, such as length × width × depth. This method avoids errors when referring to the size of certain features, such as junction boxes.

Rule 6: Be Consistent

Develop a systematic approach when working with the drawings. For instance, take measurements in a clockwise direction around a floor plan. Begin counting similar features, such as light fixtures, from the left to right or top to bottom. Whatever the procedure, it should become a standard, systematic approach.

Rule 7: Number Takeoff Sheets

If the takeoff is done by hand, always number each takeoff sheet and keep them in order. Whenever possible, tasks, groups of similar tasks, or entire sheets should be identified by their location on the drawings. For example: *Phase 2-B—Building Light Fixtures.* Numbering takeoff sheets provides a way to ensure that all sheets are included and are in order. When takeoff is done using software the organization may be less flexible (depending on the software), so add whatever definition is needed to recognize a task.

Rule 8: Define Units for Material, Work, and Assembly Items

Items or tasks that have no labor component are called *material items.* These are furnished only and will be installed under another scope of work. An example would be the rough-in boxes cast into masonry walls.

Items that have no material component and require labor only are referred to as *work items.* Examples include load testing, cleanup, and the installation of fans supplied by the heating, ventilating, and air conditioning (HVAC) contractor. All tasks or items should be labeled with a unit of measure, which will be extended to the final price.

All items that have a cost value in the estimate should be assigned a unit of measure. Most items or tasks have defined units of measure. Others are less clear and are sometimes assigned a more arbitrary unit of measure, called a *lump sum* or *LS*, most often applied to work items that are not measured or expressed in more conventional terms. A compilation or assembly of work items often uses an LS unit. When tasks are repetitive, sometimes it is easier to group them into one LS item. If you use the LS unit to incorporate multiple tasks into one item, it is important to accurately and adequately define what tasks are included, so there is no confusion.

Rule 9: Use Decimals

Decimals are preferable in the quantity takeoff in lieu of fractions, because they are faster, more precise, and easier to use with a calculator or on a computer.

Dimensions on drawings should be converted to their decimal equivalent.

For example:

A dimension of 24'-6" should be converted to 24.5'.

In calculating the area of a room that is 24'-6" × 24'-6", converting both dimensions to decimals and performing the multiplication results in an area of 600.25 square feet (SF). Always check the final units of the dimension. Adding linear dimensions results in LF or linear yards (LY). For example:

The perimeter of a rectangle that is 12' long × 13' wide is 50 LF.

Rule 10: Verify Appropriate Level of Accuracy

While accuracy is important, overaccuracy wastes time. There is an old adage in the construction business that warns, "Don't spend 10 dollars of estimating time figuring a 1-dollar item—unless there are thousands of them." Accuracy is relative to the task being taken off or estimated. Rarely is it necessary for a number to be calculated to more than two places after the decimal.

There are acceptable parameters for rounding, depending on the particular task being calculated. Most items can be rounded to the nearest full unit. In some cases, it is necessary to round to the nearest sales unit, if the balance of the sales unit has no inventory or future value. (Waste factors will be discussed in Rule 11 and in its own section following these rules.) Rounding quantities should be done when appropriate.

Rule 11: Calculate Net versus Gross Quantities

The majority of materials require an added allowance for *waste*. Waste is the difference between what is purchased and what is used. Waste is applicable to materials only, and should not be applied to labor or confused with productivity. Before waste is added, quantities are referred to as *net quantities*. After an allowance for waste has been added, quantities are referred to as *gross quantities*. (See the Accounting for Waste section later in this chapter.)

Rule 12: Check the Takeoff

If done by hand, without the use of a digitizer or computer, the quantity takeoff should be checked by another individual for overall accuracy. Ideally, it would be best to have another completely separate takeoff done as a means of checking the first. However, this may not be practical or cost effective. Quantities derived by hand should be randomly checked. Select several work items or tasks throughout the estimate and recalculate their quantities. Extensions from the takeoff quantities to the final pricing units can even be checked by a reliable clerical staff person who has minimal background in estimating. The extension of quantities involves calculations that can be checked by anyone with an understanding of simple mathematics and a calculator. While computerized takeoff reduces errors in computation, the quantities that come out are only as good as data that is input. Random checks should be done here as well. Common errors include incorrect placement of the decimal point and omission of tasks.

Rule 13: Mark up the Drawings as Bid Documents

Mark the drawings using check marks, colored pencils, or highlighters when doing takeoff by hand. Most software programs allow the estimator to color or cross-hatch the work that has already been taken off. These serve as the estimator's work papers and should be kept as a record of how quantities were derived. This aspect of the takeoff is critical for projects that are bid and then go to contract.

Rule 14: Maintain Concentration

Those who perform the takeoff and pricing require a high level of concentration in order to accurately do the job and should not be subject to interruptions from all the normal distractions of the construction office. Phone calls, frequent drop-ins by coworkers, and any type of concentration breaker are detrimental to accurate performance. Distractions or attempts to multitask are often the greatest source of error. When the takeoff must be interrupted, select a natural stopping point and mark it clearly so that when work is resumed, there is no doubt as to where you left off.

Rule 15: Organize the Documentation

Careful organization and neatness of work papers and takeoff sheets are crucial. If supporting work papers are needed (including sketches or details as to how unusual features were estimated), they should be retained and attached to the pertinent quantity sheet. Even if the takeoff is performed using a computer, there will still be work papers and notes from the document study.

All calculations should follow a logical and sequential process. Preprinted takeoff sheets and forms can be used to maintain consistency. Erasures should be neat and clean. Work papers, quantity sheets, and all components of the estimate should be maintained for a minimum of one year. Projects are sometimes abandoned for a number of reasons. Often, those same projects are restarted at a later point in time, due to changes in the economy, ownership, or need. Retaining the estimate and its various components is a reliable way to check for what has changed or remained the same over time.

A number of shortcuts can be used for the quantity takeoff. If approached logically and systematically, these techniques help to save time without sacrificing accuracy. Consistent use of accepted abbreviations saves the time of writing things out. An abbreviations list similar to the one that appears in the appendix of this book might be posted in a conspicuous place for each estimator to provide a consistent pattern of definitions for use within an office.

Rule 16: Rounding

Rounding off, or decreasing the number of significant digits, should be done only when it will not statistically affect the resulting product. The estimator must use good judgment to determine instances when rounding is appropriate. An overall 2% or 3% variation in a competitive market can often be the difference between getting and losing a job, or between profit and no profit. The estimator should establish rules for rounding to achieve a consistent level of precision. In general, it is best not to round numbers until the final stage (summary) of quantities.

The final summary is also the time to convert units of measure into standards for practical use (e.g., LF of wire to CLF units). This is done to keep the numerical value of the unit cost manageable. The installation of type THW #14 wire requires .00615 labor-hours per linear foot, but this information is expressed as .615 LH/CLF (labor-hours per 100 linear feet).

ACCOUNTING FOR WASTE

Quantities derived during the takeoff process are often not the same quantities that are purchased when the work is actually in process. For example, it may be determined that the actual length of a conduit run is 97 LF. However, the conduit is sold in bundles of 100 LF. Since only full bundles are sold, 1 bundle must be included in the takeoff. The difference between what to include in the takeoff and estimate and what is actually installed is called *waste*. As mentioned earlier, the material quantities before waste is added are called *net* quantities, and *gross* quantities after waste has been added. Pricing is done at the gross quantities level, not on net quantities.

Waste may need to be added for any of three primary reasons:

- To adjust to the standard sales unit.
- As anticipated waste resulting from handling.
- To achieve a specific assembly lap or coverage.

Adjusting for Standard Sales Units

Materials often go through some on-site modification. The classic example is wire. Wire, in spools or coils, is sold at predetermined lengths. The wire is delivered to the site and then cut to exact lengths for the specific application. It is purchased by converting the total length required to the number of spools or coils as close to the in-place quantity as possible to minimize waste. The remaining or residual wire left over after the installation, sometimes referred to as *fall-off*, is the waste. It may have a value on another project, but still needs to be accounted for in the estimate, because it is paid for at the time of purchase.

It is important to be attentive to other types of materials with similar waste requirements. Any material with a standard sales unit larger than needed for the task qualifies as having a waste component. Construction materials sold in lengths, rolls, bundles, boxes, and sheets, and fluids sold in gallons, drums, or barrels, should be reviewed for waste.

Waste Resulting from Handling/Storage

Waste can occur as a result of handling or placement, which is fairly common. Even with careful planning and execution of a task, waste will occur. Often, these materials are distributed by equipment and can be damaged in the process. Generally, the more the materials are handled, the more waste can be expected. Materials that are stored for excessive periods can also be damaged and may need to be replaced. Weather, humidity, and moving materials from one staging area to another can often result in losses.

Waste Required for Lap

Often, additional materials are required to satisfy a specific *lap* in order to maintain continuity of a particular feature. While lap is less common in the electrical trade than in others, it still does occur. Allowing for lap does not meet the strict definition of waste, since the material is actually used in the project, but lap requires additional materials, so the same principle applies.

OTHER FACTORS THAT AFFECT QUANTITIES
Economy of Scale

In addition to the specific examples previously noted, there are other considerations that, while not specifically considered waste, have an impact of the amount of materials included within the takeoff. Price breaks based on total quantities should also be taken into account, referred to as *economy of scale*. This is a simple economic principle that can be defined for our purposes as securing a better unit price for a large quantity of a material purchased.

Once the takeoff has been completed, the next step is to start the pricing phase of the estimate. In order to accurately apply unit prices to quantities, one must understand the different types of costs associated with the unit price system.

5 | Pricing the Estimate

When the quantities have been derived and the takeoff is complete, the next phase of the estimate can begin. This phase involves the addition of prices and multiplication of quantities by the unit price to arrive at the total cost for a task. It is aptly named the *pricing phase*.

It is at this point in the process that the estimator must select one of the methods for pricing described in Chapter 1, "Components of an Estimate." In many electrical contracting companies, it is less the estimator's decision than a matter of company policy as to the method used. In unit price estimating, the unit costs most commonly applied are *bare* or *unburdened* costs. The burden is then added at the summary stage. This is also true for materials, labor, and equipment. In the following paragraphs, we will demonstrate by using the most common method.

A majority of electrical estimators calculate the cost of labor by labor-hours. When the total labor-hours are determined, it is necessary to multiply these figures by the appropriate hourly rate before adding markups. The extended labor quantities are summarized as total labor-hours to which burdens are applied. Material quantities extended to costs are represented as dollars. Items such as overhead and profit are usually added to the total direct costs at the estimate summary.

SOURCES OF COST DATA

One of the most difficult aspects of the estimator's job is determining accurate and reliable cost data to be used in the preparation of an estimate. Sources for such data are varied, but can be categorized in terms of their reliability. The most reliable of any labor cost information is the accurate, up-to-date, well-kept records of completed work by the estimator's own company. There is no more accurate cost of the labor component for a particular task than the *actual* cost to the contractor of that task from another recent job. This is referred to as *historical* cost data, and in the marketplace of competitively bid electrical projects, there is no substitute. Historical cost data represents the completed costs and productivities by the personnel that are performed and supervised the work on the contractor's payroll. It should be added that costs from a previous project may require modification to represent the new tasks to be estimated. The use of historical data is not merely a mechanical application of numbers, but requires a significant measure of experience and judgment.

The best sources of material prices are up-to-date quotations by vendors and suppliers that are specific to the project being estimated. This is referred to as *contemporaneous* pricing and is the predominant methodology used. The same is true for equipment rental pricing. A point of clarification is in order; in estimating practice, the term *material* refers to all items that will be permanently installed in the project. This includes items that are sometimes termed *equipment,* such as panelboards, motor starters, and transformers. Equipment in estimating practice are aids to the construction process that are temporary in nature and are not incorporated in the final work, such as scaffolding, lifts, temporary generators, bucket trucks, and so on.

This same methodology is applied to any subcontractor pricing provided to the electrical subcontractor. Subcontractor quotations are contemporaneous in that the pricing is specific to *that* project. The subcontractor considers not only the materials and physical work, but the terms and conditions that make the project unique, such as the schedule, season of the year, labor requirements, crew sizes, milestones, applicable taxes, and insurance rates at the time estimated. This is essential for accurate pricing of subcontracted scopes of work.

All price quotations from vendors, suppliers, equipment rental companies, or subcontractors must be obtained in writing. In today's world of near-instant electronic communication, there is no exception to this rule. Qualifications and exclusions should be clearly stated in the proposal. The scope or items quoted should be checked to be sure that they are complete and as specified. This is called *qualifying* a bid. One way to ensure that these requirements are met is to request that the quotation satisfy the contract documents or that the bid is *as per plans and specifications.* This means that the quote fully satisfies the terms, conditions and scope as represented by the bid documents.

Another, although less secure, method is to devise a form, called a *request for proposal* or *RFP* and define the scope that the bidding party will base their price on. The RFP can define the scope and all of the appropriate terms and conditions that must be considered when preparing the estimate. The reduced security results in the added layer of responsibility for the electrical contractor's estimator in capturing all of the work required in the scope correctly.

For the estimator who has no cost records for a particular item and is unable to obtain contemporaneous pricing, then the next most reliable source of price information is published cost data. A current unit price cost book such as *Means Electrical Cost Data*™ published by the RSMeans Company is an essential asset. In addition to being a source of costs, unit price books can also be useful as a reference or cross-check for verifying costs obtained elsewhere or for crew size and recognized productivities.

Lacking cost information from any of the previously mentioned sources, the estimator may have to rely on experience and personal knowledge of the field to develop costs.

Regardless of the source of cost information used, the system and sequence of pricing should be the same as that used for the quantity takeoff. This approach must be consistent and should continue until the estimate is complete.

COST CATEGORIES

All project costs can be separated into two main categories: *direct* and *indirect* costs. Direct costs are those costs that are directly related to a single project and are necessary to execute that work of that one project. Indirect costs are relative to the cost of operating the electrical contracting company and are not attributable to only one project, but are shared or *absorbed* by all the ongoing projects.

A simple question can almost always define which category a cost will fall into: *if I did not have the project, would I still have the cost?* If the answer is yes, then the cost is an indirect cost. If the answer is no, then the cost is a direct cost. This question holds true regardless of the accounting method employed by the contractor. For example, the cost of the project manager can be a direct cost if his or her salary and benefits are directly attributable to a project or can be indirect if the project manager is considered part of home office overhead.

TYPES OF COSTS

Within each cost category in an estimate, there are types of costs. Types of direct costs include materials, labor, equipment, subcontractors, and project overhead. Some of the project overhead costs can be fixed, and some vary with time. An in-depth discussion of direct costs will be conducted in Chapter 6, "Direct Costs."

Indirect cost types include costs of operating a company such as the salaries of principals, rent on the office/shop, clerical and support salaries and benefits, heat and light bills for the office, advertising and promotion, tax preparation, and other types of business management costs. To recover these costs, a small amount of each is added to the bids as a percentage of the cost of the work. These costs are tracked weekly or biweekly and adjusted as dictated by the financial health and workload of the company. Adjustments can include the downsizing of the cost or the outright elimination of the cost.

A clear understanding of direct and indirect costs is a fundamental part of pricing the estimate. A thorough discussion of each of these types of cost will be presented in Chapter 6, "Direct Costs," and Chapter 7, "Indirect Costs."

PROFIT

Profit is the reason for doing business. It is the end result of a project done right and/or the reward for risks taken. While profit is not specifically a cost, it does need to be captured in the calculation of the billing rate. Without it, the project would be considered a failure. Profit is most often assigned as a percentage of all costs of the work. However, it can be assigned as a lump sum or stipulated fee. Calculating profit is based on careful consideration of a host of factors—some less tangible than others. A complete discussion of the assignment of profit will be covered in Chapter 7, "Indirect Costs."

6 | Direct Costs

Direct costs can be defined as those costs necessary for the completion of the project—in other words, the hard costs. Material, labor, and equipment are among the more obvious items in this category. While subcontract costs include the overhead and profit (indirect costs) of the subcontractor, they are considered to be direct costs to the prime contractor. Also included are certain project overhead costs for items that are necessary for construction. These are called *project overhead costs* or the general requirements of the project. Examples are a storage trailer, tools, temporary power, lighting, lifts, and so on. Sales tax and bonds are additional direct costs, since they may be contractual requirements of the project.

MATERIAL

As noted in the previous chapter, accurate material pricing for competitively bid electrical projects must be contemporaneous. Ideally, the vendor (or supplier) should have access to the plans and specifications for verification of quantities and specified products. *Note:* The author makes the distinction between vendor and supplier as follows: a vendor furnishes a definitive, finite scope of materials to the project, for example a lighting package or the switchgear. A supplier is defined as the entity that provides more common electrical materials such as wire, conduit, device boxes, and so on. The supplier typically will not provide a fixed price with a fixed quantity of these items, but instead will continue to supply them to the project as long as the monthly bill is paid. In contrast, the vendor has a fixed quantity for a fixed price and completes their purchase order when the materials quoted have been delivered. On many projects they are the same entity.

Material pricing appears relatively simple and straightforward. There are, however, certain considerations that the estimator must address when analyzing material quotes. These considerations often extend beyond just the cost of material. One of these considerations is the reputation of the vendor. This can be a significant factor. Prior experience with the vendor is usually the milestone for making such decisions. The estimator must be satisfied that the scope bid is comprehensive and based on the materials specified. When the vendor is invoking the *or equal clause* in a product specification, the estimator must be comfortable that the substituted product is demonstrably equivalent to the product specified. Failure to demonstrate equality may result in the electrical contractor's having to provide the specified product even though the cost was not included in the estimate. Other factors include whether the vendor is qualified to satisfy the needs of the contract, both financially and as a recognized distributor of the products required. More than one contractor has been damaged by the poor performance of a vendor.

Estimators may choose not to rely on a lower price from an unknown vendor, but will instead use a slightly higher price from a known, reliable vendor with which they have a history.

There are many other questions that the estimator should ask: How long is the price guaranteed? Is there a timed escalation clause for the product? Does the price include delivery charges or sales tax, if required? Are guarantees and warranties in compliance with the specification requirements?

The estimator must be sure that the quotation or obtained price is for the materials as per plans and specifications. In some cases, substitutions can substantially lower the cost of a project. Note also that many specification packages will require that *catalog cuts* or product literature be submitted for certain materials as part of the substitution proposal. The vendor proposing the substitution should be more than willing to prepare the product literature needed to secure approval for the substitution.

When the estimator has received material quotations, there are still other considerations that should influence the final selection of a vendor. *Lead time* for the products, or the amount of time between order and delivery, should always be a consideration. It is pointless how competitive or low a quote is if the material cannot be delivered to the job site on time to comply with the schedule. Many vendors will make unrealistic promises for delivery to close the deal. The delivery date should be in writing and part of the terms and conditions of the sale.

The estimator should also determine whether there are any unusual payment terms. Payment terms outside of the normal 30-day business cycle can impact the decision process. Cash flow for a small electrical contractor can be severely affected if a large material purchase is outside normal terms.

There is another factor that can impact material pricing: tax. Tax can be as simple as a fixed percentage of the materials cost for sales tax. It can be a residence or use tax on the entire contract amount or no tax at all. Taxes will vary state by state, and even county by county. There are locations that have a city tax as well as a state tax. The estimator must become familiar with the applicable tax rules of the location of the project. On projects with single-digit profit margins, omitting a 5% or 6% sales tax can be a death knell to the project.

Some taxpayer- and government-funded projects are exempt from sales tax, so there may be no need to add it. However, most residential and commercial projects are subject to a sales tax on materials.

LABOR

Labor has been and will always be the most important component in the estimate. It is the wild card! To determine the installation cost for each item of construction, the estimator must know two pieces of information: first, the labor cost or *labor rate* of the tradesperson, and second, how much time a worker will need to complete a given unit of the task. This is more commonly referred to as *productivity*.

Billing Rate

Calculating the cost of labor unit prices is more difficult than for material unit prices. First, the correct wage rate must be calculated. There are several modifiers that must be added to the cost of the actual wage rate. These include state and federal taxes on wages, insurances, and benefits.

We should start by defining some basic terms. The term *wage rate* for this discussion refers to the "in the envelope pay," or the agreed-upon wage between the employee and employer. Once the wage rate has been modified by all taxes, insurance, and overhead components, the rate is referred to as the *burdened labor cost*. After profit has been added, the result is the full value of the labor-hour, referred to as the *billing rate* or labor rate. This is the rate that would be charged for labor in a time and material application. The following is a review of the components that make up the billing rate and how those costs are calculated. It should be noted that the billing rate is for time and materials work or *service billing*. As will be demonstrated in Chapter 8, "The Unit Price, Project Overhead Summary, and Estimate Summary Sheets," the labor rate used for work that the contractor will self-perform will be exclusive of overhead and profit. In this methodology, the overhead and profit will be added in the Estimate summary.

Wage Rate

The wage rate, or the hourly rate of pay on which the employee's paycheck is calculated, is regulated by the agreement between the employee and the employer, a collective bargaining agreement, or by a prevailing wage rate in the case of taxpayer-funded work. For example, an electrician hired at $30 per hour would gross $1,200 in wages for a 40-hour work week. From this amount, taxes would be deducted, and the employee would receive a net paycheck.

Benefits

Many employers provide *benefits* in the form of medical and dental policies, vacation pay, paid sick days, retirement package contributions, annuities, or a variety of other compensatory benefits. For union employees the benefit package is regulated by the collective bargaining agreement. For the open shop employee, benefits may be what is negotiated in the hiring interview or what is offered by the employer. In either application, the costs of these benefits are included within the calculation of the billing rate. These benefits represent a cost to the employer that must be recovered, which can be broken down into a percentage of the hourly rate and then extended to a dollar amount per hour. This dollar amount is then added to the wage rate.

State and Federal Taxes and Insurance

All states and the federal government apply a tax to wages to provide a source for unemployment and social security benefits. The *Federal Unemployment Tax Act* or *FUTA* and *State Unemployment Tax Act* or *SUTA* are taxes paid by the employer based on the dollars earned by the employee. It is irrespective of the trade. They are levied on the employee's gross taxable wages up to a maximum called a *cap*. Once the employee earns more than the cap, the tax stops. The SUTA tax varies from state to state and will fluctuate depending on the employer's experience, which is measured by the number of employees that the employer has laid off. The fewer employees claiming benefits, the less the employer's contribution. As the number of employees claiming benefits increases, the amount of the employer's contribution also increases. The SUTA tax can vary with a number of factors based on the employer's experience in the workforce, number of workers, and the type of work. There is typically a cap on the wages, as with the FUTA tax.

The *Federal Insurance Contributions Act* or *FICA* is Social Security and Medicaid taxes, which are provided by the federal government, though more appropriately considered insurances than taxes. The employee is taxed at the rate of 7.65% of gross taxable wages, deducted from the employee's paycheck. Additionally, the employer makes a matching contribution of the same percentage. It is this percentage that is recouped in the wage calculation.

Workers' Compensation Insurance

Another type of insurance that protects employees is *workers' compensation*, provided by the employer in the event of injury, disability, or death occurring in the workplace. In most states, this is a compulsory insurance. Workers' comp, as it is commonly referred to, is different for each classification of worker or trade, and is based on a percentage of the gross nonpremium wages of the employee. (In other words, overtime or premium wages do not affect the insurance premium.) Insurance premiums for individual companies performing the same work will vary by experience. Premium rates are based on the employer's number of days without a serious accident on the job site. A safe work environment translates to lower workers' compensation premiums based on the *experience modifier*. Thus, the insurer rewards the employer for promoting a safe work environment with reduced insurance premiums. Lower premiums are reflected in lower labor costs, thereby enabling the contractor to be more competitive in bidding.

General Liability Insurance

General liability insurance protects the project and any adjacent property from damage. This insurance compensates the contractor and, ultimately, the owner for damage by parties under the control of the general contractor, including employees and subcontractors. General liability premiums are again rated on the contractor's success in reducing risk. They are calculated based on two factors: employee wages and subcontracted work. For most prime contractors who subcontract more than 70% of their contract work, the largest portion of the premium is based on the costs incurred by those parties under subcontract. However, there is still a portion of the premiums that is based on gross employee wages. Percentages will vary with the limits of the policy, size of the work, performance history of the contractor, and any regulations mandated by the state or insurer.

Again, for parties without the historical data to draw from, RSMeans produces *Means Electrical Cost Data*, which provides national averages for an electrician's labor rate, fixed taxes, and insurance. Some states also apply sales tax to labor.

Labor Productivity

Productivity can best be described as the rate at which work is produced by an individual or crew per unit of time. The unit of time most often used is the *workday*. A workday is considered an eight-hour period, over which productivity will fluctuate up or down, depending on breaks, learning curves, starts and stops, and material handling of the task. (The hour is too short a period to provide an accurate measurement.) Productivity on the same task can vary when performed by two different crews, or during different times of the day. It can also vary with extremes in weather or temperature.

When the work produced is measured over the course of the whole workday, it takes into account all of the normal fluctuations. It is fairly easy to take a "snapshot" at the beginning of the day and again at the end of the day and compare the difference. This is how productivity is measured.

There are many factors to consider when determining the expected productivity of a task. For example, consider two separate conduit runs with the same total linear footage; one run is 8' above the floor, and the other is 28' above the floor. All things being equal (personnel, crew size, wage rate, pipe size, etc.), with the only difference being the height above the floor, most electrical construction professionals, even novices, would agree that the 28' height reduces mobility, thereby making the installation process slower and more difficult. This is translated into reduced productivity and a higher unit cost for labor (less units completed per hour). This concept of varying productivity exists throughout the construction industry on virtually every task. So in order to accurately estimate the labor cost of a task, the estimator has to understand the *context* (8' or 28' above the floor) in which the work is being performed in order to determine how the work will be done and the appropriate crew to do the work. Having experience as a tradesperson is always helpful understanding the requirements of the task being estimated.

The best measure of an individual or crew's productivity is its own historical data. Job records for payroll and labor-hours to complete a specific task offer the best guidelines for predicting future performance. This is where the adage *"Past performance is a good indicator of future performance"* is demonstrated. Experienced estimators know that while this is no guarantee that future performance will be the same, it provides a logical model that serves as a basis for a defendable labor estimate. Specific adjustments can be made based on the needs of the individual project or its unique conditions. For example, in the preceding conduit illustration, the productivity data for the 8' height and the 28' height might be used to extrapolate the productivity for a 12′ height conduit run on a project being bid. Wage rates are known when bidding a project, but productivity may require more work to accurately determine.

Estimators working for electrical contractors should have well documented productivities for routine tasks on projects the company has performed. For tasks that are not within the realm of the company's documented data, published data can be used to compare productivities and crew composition. Cost data books, such as *Means Electrical Cost Data*, provide national average productivities for thousands of tasks. These productivities can be used in the absence of historical data or as a comparison check to a company's own data.

Individual and Crew Tasks

All tasks can be categorized for estimating purposes into either an individual task or a crew task. An individual task is one performed by a single individual such as an electrician. Performance and productivity is measured by the single output of the individual performing the work. As an example, consider an electrician installing duplex receptacles and the cover plates. This is clearly an individual task. The performance of the individual electrician can be easily measured by counting the quantity of devices installed in a single day.

After several days of installing duplex receptacles, an average productivity could be established, and a model could be created for future installations. The cost of the electrician per day could then be divided by the number of devices installed to establish a baseline cost model per device. Adjustments could be made to this cost model to predict the productivity for different conditions than the one used to create the cost model.

An example of a cost model for the electrician installing duplex receptacles is: 8 electrician hours × $78.40 per hour = $627.20 per day. Over multiple days, it was determined that the electrician installed an average of 27 duplex receptacles and plates per day. From the data, it could be concluded that approximately 0.296 electrician-hours are needed to install the device and plate and at a cost of $23.23 per device.

Crew tasks can be more challenging to estimate in terms of productivity. They combine the performance of multiple individuals, each of whom performs a specific function to complete the overall task. The measurement of productivity is a by-product of how well the team performs together. Crew tasks have a different dynamic—efficiency as a team. For example, a crew pulling feeders might consist of two journeyman electricians (one of whom is a foreman) and two apprentices. Each member of the crew has a specific function that contributes to the overall performance of the crew. The foreman directs the crew, manages the work, and aspires to a steady, predictable productivity. The foreman may produce less as a result of other duties, but still contributes. The apprentices on the crew provide support services to the journeymen. While they do not have the same level of responsibility as the journeyman, their duties are essential to successful production of the crew.

Each crew member's performance relies on the other members to perform their work efficiently. Adding or removing a single individual changes the crew

Table 6.1   Elevated Equipment Installation Factors

Installation Height Above Floor	Labor Adjustments
10′ to 15′	+ 15%
15′ to 25′	+ 30%
Over 25′	+ 35%

For the installation of electrical equipment elevated more than 10′ above the floor, the labor costs should be adjusted to allow for the added complexity. This table lists suggested adjustments.

Source: Reprinted from *RSMeans Estimating Handbook*, Third Edition, Wiley.

composition and the productivity. Selecting the most efficient crew is critical to successfully estimating. Historical data plays a key role in determining the crew size for maximum efficiency and productivity. Experience is also a major contributing factor that can be used to temper the crew composition.

Productivity can also be affected by conditions such as its height above the floor. It is a recognized fact that work performed above 10′ off the floor is less productive and generally costs more in labor. Table 6.1 provides some guidance for increased labor costs for work performed over 10′.

EQUIPMENT

Construction equipment has become more important, not only because of the incentive to reduce labor costs, but also as a response to new, high-technology construction methods and materials. As a result, equipment costs represent an increasing percentage of total project costs in construction. Estimators must carefully address the issue of equipment and related expenses. Equipment costs can be divided into the two following categories:

1. **Rental, lease, or ownership costs.** These costs maybe determined based on hourly, daily, weekly, monthly, or annual increments. These fees or payments buy only the "right" to use the equipment (i.e., exclusive of operating costs).
2. **Operating costs.** Once the "right" of use is obtained, costs are incurred for actual use or operation. These costs may include fuel, lubrication, maintenance, and parts.

Equipment costs, as described earlier, do not include the labor expense of operators. This is calculated separately as described in the previous section on labor and added to the equipment cost.

Equipment ownership costs apply to both leased and owned equipment. The operating costs of equipment, whether rented, leased, or owned, are again most reliable when they originate from historical data. These operating costs consist of fuel, lubrication, expendable parts (such as blades or bits), and regular maintenance costs. For estimating purposes, the equipment ownership and operating costs should be listed separately. In this way, the decision to rent, lease, or purchase can be decided project by project.

Another consideration is portal-to-portal transportation costs and setup and breakdown costs for larger equipment. This is called *mobilization* and *demobilization*. It is the time and cost associated with the setup of a piece of equipment before it can be used. A crane is a classic example. It is charged from the time it leaves the yard until it returns to the yard (portal-to-portal). When it gets to the site, it often has some assembly and stabilization that needs to be done before it can start hoisting.

There are two commonly used methods for including equipment costs in a construction estimate. The first and most common is to include the equipment as a part of the construction task for which it is used. In this case, costs are included in each line item as a separate unit price. The advantage of this method is that costs are allocated to the division or task that actually incurs the expense. As a result, more accurate records can be kept for each installed component. The estimator must consider that although piece of equipment is required for only 3-hours to complete a task, there may be a minimum rental that exceeds the needed period. The same may be true of the operator's labor cost.

The second, less common method, for including equipment costs in the estimate is to keep all such costs separate and to include them as a part of project overhead. The advantage of this method is that all equipment costs are grouped together, and that machines used for several tasks are included (without duplication). One disadvantage is that for future estimating purposes, equipment costs will be known only on a per-job basis and not per installed unit. This method is most often used when the entire project consists of repetition of the same task, for example, the setting of streetlight poles.

Whichever method is used, the estimator must be consistent and must be sure that all equipment costs are included but not duplicated. The estimating method should be the same as that chosen for cost monitoring and accounting. In this way, the data will be available both for monitoring the project's costs and for bidding future projects.

Some taxpayer- and government-funded projects are exempt from sales tax, so there may be no need to add it to rental equipment. However, most residential and commercial projects are subject to a sales tax on equipment that is rented.

A final word of caution about equipment is to consider its age and reliability. If an older item, such as a bucket truck, needs frequent repair, it may cost far more in lost labor-hours to the project than is reflected in its calculated equipment cost rate.

SUBCONTRACTORS

Subcontractors are, by definition, independent contractors. They are an indispensable part of the construction process and perform an increasingly larger percentage of work on construction projects. Subcontractors "hire and fire" their own labor, purchase their own materials, run their own equipment, and, quite often, define their success entirely on their own performance. Subcontractors provide their own insurances, both general liability and Workers' Compensation. Their bids include overhead and profit. They are held responsible for their actions and are required to pay their own taxes. This is the very essence of sub-contracting work—sharing risk.

By virtually every test, subcontractors are considered independent of the prime (electrical) contractor. Subcontractors are bound by agreement to prime contractors to execute a portion of the work. This portion of work is most often well defined but of limited scope, with a stipulated sum as the basis of compensation. This sum is a piece of the overall or total amount that comprises the contract value between the prime contractor and the party to which they are bound (general contractor [GC] or owner).

Typical agreements between GCs (or owners) and electrical contractors do not explicitly recognize the electrical contractor's subcontractors. The agreement assigns responsibility for subcontractors to the prime contractor, who is the only signatory party with the GC (or owner). However, prime contractors realize that there is a price to pay for managing and administering the work of subcontractors—overhead and profit. The prime (electrical) contractor must include overhead and profit on the subcontractor's work. This will be discussed in Chapter 8, "The Unit Price, Project Overhead Summary, and Estimate Summary Sheets."

When subcontractors are included in the prime contractor's bid, quotations should be solicited and analyzed in the same way as material quotes. A primary concern is that the bid covers the work as per plans and specifications, and that all appropriate work alternates and allowances are included. Any exclusions should be clearly stated and explained. If the bid is received verbally, it must be followed up with a written proposal to be considered legitimate. Any unique terms or conditions that differ from the bid documents must be noted and evaluated prior to inclusion in the prime's bid. Such requirements could affect (restrict or enhance) the normal progress of the project and therefore should be known in advance.

The estimator should note how long the subcontract bid will be honored. This time period usually varies from 30 to 90 days and is often included as a condition in complete bids.

The estimator should question and verify the bonding capability and capacity of unfamiliar subcontractors. This includes the name of the surety, since some bonds are not worth the paper they are written on. Taking such action may be necessary when bidding in a new location. Other than word of mouth, these inquiries may be the only way to confirm subcontractor reliability.

One final note concerning subcontractors. Our very own Internal Revenue Service (IRS) has specific criteria for defining a subcontractor. These criteria have survived many a legal challenge. All parties should be aware of applying the term *subcontractors* to what otherwise could be considered a thinly veiled attempt at avoiding the payment of insurance and taxes on wages. The common practice in the residential construction industry of hiring individuals, paying them as subcontractors by issuing a "1099" at the end of the year, and not insuring them with Workers' Compensation and liability insurance may not survive the challenge and could result in fines or penalties being assessed to both parties. If you are unsure of the rules, seek professional guidance!

PROJECT OVERHEAD

The technical portion of the specifications is based on the 50 divisions of the CSI MasterFormat. The first division is appropriately named Division 1—General Requirements and is the only topic within the General Requirements subgroup. This division deals primarily with:

- Project overhead requirements.
- The basis for administering the project, as defined by the General Conditions of the Contract for Construction.
- Administrative requirements.

While the General Conditions of the Contract and the General Requirements of Division 1 stand alone as two separate documents, they are related. The General Requirements are important to the estimating process because they provide the information needed to assign a monetary value to the project overhead items.

General Requirements include items such as:

- Temporary facilities and controls
- Project meetings
- Reference standards and definitions
- Submittals
- Testing requirements
- Project closeout
- Project record documents
- Quality control
- Commissioning
- Project phasing
- Progress schedule (critical path method [CPM])

Division 1 also identifies the contractor's special contractual obligations that have an associated cost and must be accounted for in the bidding process. This category includes development and provision of:

- Unit Prices
- Alternates
- Allowances

While the General Requirements are the first division of the CSI MasterFormat, they are often the last to be estimated. This is because most General Requirements items require a complete understanding of the entire project, which is not possible until after the estimator has become familiar with the contract documents, performed the takeoff of electrical and related work, made a site visit (if applicable), and estimated the majority of the project.

One of the most important parts of Division 1 is project overhead or *direct overhead*, often called the nonproduction costs or the cost of supporting the production activities. Direct overhead costs are directly related to one individual project. These include a wide variety of costs such as temporary facilities, trailer rental, supervision, telephone usage, dumpsters, and electrical power usage. All are costs that are directly related to one specific project. They can be broken down further for estimating purposes into time-sensitive costs and fixed costs.

Time-Sensitive Costs

Time-sensitive costs are items whose price is driven by time or are schedule driven. The longer the particular item is on site, the higher the cost. A good example would be a storage trailer. Most contractors rent trailers, and since there is a monthly fee for rental, the cost is a function of time, or time sensitive. The longer the project goes on, the higher the accrued cost for the trailer rental. There are numerous other examples of time-sensitive costs that can be found in project overhead, temporary services such as lighting and power distribution costs. To accurately assign a dollar value to these costs, a schedule should be developed to determine how and when they apply during the term of the project.

Initial schedules for determining time-sensitive costs tend to be rudimentary and develop with more information as the estimate proceeds. It is not uncommon for a project schedule to evolve through multiple generations before it is considered sufficient for use in the estimate (The development and evolution of schedules for estimating purposes will be discussed in detail in Chapter 9, "Prebid Scheduling.")

Fixed Overhead Costs

Direct overhead costs that are not affected by time are classified as *fixed* project overhead costs. Examples include electrical permit fees, records documents, engineering design fees, testing services fees, and so forth. In most circumstances, there is a single occurrence for each of these costs, independent of the project schedule.

Project overhead, like all other direct costs, can be separated into material, labor, and equipment components.

Applying Prices to General Requirements

Applying prices to fixed-cost items is fairly straightforward. Prices for time-sensitive, or variable, cost items can be determined using the list of needed items, together with the schedule developed or provided by others. General Requirements costs, such as supervision, are reasonably clear. The General Requirements stipulate that the project is required to have competent supervision by the electrical contractor, defined as supervision licensed by the authority having jurisdiction, with a particular education and experience level in the type of project. The General Requirements might mandate that the supervisor be on-site during all times that the work is in progress, including weekends and holidays. These statements help to determine what level of supervision is required, and the associated salary and benefit costs. At that point, it becomes a simple unit price multiplied by the duration.

Other time-sensitive costs may be required only intermittently or for specific periods within the overall project duration. For example, the General Requirements may stipulate that the electrical contractor provide a storage trailer on-site until the building is closed in, and then materials can be stored in the building. The estimator would then calculate by the schedule how many weeks or months until the building will be closed in, and an indoor storage area can be set up. The period (in weeks or months) from the start of the project until the date the building could accommodate storage would be determined and then multiplied by the appropriate rate (per week or month).

Costs for General Requirements categories such as project meetings and project closeout are judgment calls and almost exclusively experience based. It is difficult to predict accurately how long site meetings will take before the project starts. Any prediction is based purely on past experience, and that is still no guarantee. Possible considerations in determining costs for meetings include level of architect or engineer involvement (the more involved, the longer the meetings), level of document development (the more complete, the shorter the meeting time), and contractual requirements. Many contracts mandate a weekly meeting.

It should be noted that when the electrical contractor is a subcontractor to the GC, the Division 1—General Requirements section still applies to the electrical subcontractor. This pass-through of responsibility is called the *flow down* provision of the technical specifications. The estimator is directed to read Division 1 thoroughly for a clear understanding of what is to be provided by the GC and what is to be provided by the electrical subcontractor. For example; scaffolding may be the responsibility of the individual subcontractor, or it may be the responsibility of the GC to provide it for all subcontractors. Another example is that it is not uncommon for the electrical subcontractor to be responsible for installation and removal of temporary power and lighting on a project, as well as the maintenance of the temporary power and lighting, that is, relocating power sources and changing lightbulbs as the work progresses. This same flow down provision may be applicable to any subcontractor to the electrical contractor. Qualify sub-bids carefully!

BONDS

Although bonds are a direct cost, they are priced and based on the total bid amount. For that reason, they are generally calculated after indirect costs have been added and appear in the estimate summary. Bonding requirements for a project will be specified in Division 1—General Requirements, and will be included in the construction contract. Various types of bonds may be required. The following are a few common types:

- **Bid Bond.** A form of bid security executed by the bidder or principal and by a surety to guarantee that the bidder will enter into a contract within a specified time and furnish any required Performance or Labor and Material Payment Bonds. The cost of a bid bond, depending on the surety, can be negligible and is sometimes considered an indirect overhead cost.
- **Completion Bond.** The guarantee by a surety that the construction contract will be completed and that it will be clear of all liens and encumbrances.
- **Labor and Material Payment Bond.** The guarantee by a surety (guarantor) to the owner (obligee) that the contractor (principal) will pay for all labor and materials used in the performance of the contract as per the construction documents. The claimants under the bond are those having direct contracts with the contractor or any subcontractor.

- **Performance Bond.** (1) A guarantee that a contractor will perform a job according to the terms of the contract. (2) A bond of the contractor in which a surety guarantees to the owner that the work will be performed in accordance with the contract documents. Except where prohibited by statute, the Performance Bond is frequently combined with the Labor and Material Payment Bond and colloquially referred to as a Performance and Payment or P&P Bond.

The costs of a bond will vary with a variety of conditions of the principal (contractor). The estimator should check with senior management for the rate.

Subcontractor Bonds

For many prime contractors it is policy to require their subcontractors be bonded for a substantial subcontract. The term *substantial* may vary with individual contractors, but a general rule is that when the subcontractor's contract amount exceeds 10% of the prime's contract, secure a performance and payment bond! The reason is the impact that a sub at that contract sum could have should they default. With profit margins well below 10% in a sluggish economy, a default of a subcontractor has the potential to prevent the prime from satisfying their contract. Bonds on subs can vary from 1.5% to 2.5% depending on a variety of factors. In a prime/subcontractor relationship the principal is the sub and the prime is the obligee who retains the benefit of the bond. Again, it is form of managing the risk inherent between the sub and the prime contractor.

A final thought on bonding. The estimator should note that some bonds are not worth the paper they are written on and that claims against the surety may never be satisfied. The U.S. Department of the Treasury provides a listing and rating of sureties on their web site to provide information to perspective obligees or principals called *Circular 570*. The site www.fms.treas.gov/c570/c570_a-z.html is extremely helpful in determining whose bonds have value and whose do not.

Sample Calculation

Following is a sample calculation to illustrate the billing rate for an electrician from the wage rate, using hypothetical rates for insurance, taxes, and other markups:

Electrician wage rate	$30.00	per hour
Benefit package	10.55	
FICA (7.65% on wage)	2.30	
FUTA (0.8% on wage)	0.24	
SUTA (7% on wage)	2.10	
Workers' comp at electrician rate of 11.05% (on wage)	3.45	
General liability insurance at a rate of 0.82% (on wage)	0.25	
Subtotal	$48.89	
Indirect overhead at 10% of all costs	4.89	
Subtotal	$53.78	Burdened labor cost
Profit at 10% of all costs	5.38	
Billing rate	**$59.16**	**per hour**

7 | Indirect Costs

ndirect costs can be defined simply as the costs of operating a business. These expenses are sometimes referred to as main office or home office overhead. Indirect costs may include certain fixed, or known, expenses as well as costs that can be variable and only approximated. Home office overhead for a company that is well established and has created a market niche, may vary only slightly year to year. For new, growing companies, home office overhead can be volatile and change monthly. It make take years to stabilize. While not meeting the strict standard as a cost, profit and contingencies are called a cost as they are a cost to the client that pays the bill. Profit and contingencies are more subjective and can often be determined by less tangible formulas than other indirect costs. They are based on the judgment and discretion of the person(s) responsible for the company's growth and business plan.

Home office overhead, contingencies, and the assignment of profit are often the deciding factor in winning a bid. The rationalization of that statement is that if direct costs are relatively close from bidder to bidder, and they should be, then winning comes down to the indirect costs applied in the Estimate Summary. For the purpose of this chapter, *indirect, home office*, and *main office* overhead will be considered synonymous terms.

The direct costs of a project must be itemized, tabulated, and totaled before the indirect costs can be applied to the estimate. Indirect costs can be classified into one of the following groups:

- Home office overhead
- Contingencies
- Profit

HOME OFFICE OVERHEAD

Home office overhead, or the cost of doing business, is costs related to the operation of the main (home) office and its staff. These costs are associated with maintaining a business, but not directly attributable to any one specific project.

Indirect or home office overhead includes, but is not limited to expenses such as:

- Corporate officers' salaries and benefits, insurance, and taxes.
- Rent or mortgage for the home office or shop.
- Monthly telephone, fax, or Internet charges.
- Clerical and office associates' salaries, insurance, and taxes.
- Corporate vehicles—lease or purchase.
- Interest expense from loans.
- Vehicular insurance, fuel usage, and maintenance costs.
- Estimator's salary and benefits.
- In-house bookkeeping and accounting costs.
- Payroll services.
- Heating, electricity, and maintenance of the main office.
- Real estate taxes on office and excise tax on vehicles.
- Hardware and software updates, training, and expenses.
- Advertising and promotional materials and costs.
- Association membership's dues.
- Profit sharing or retirement contributions.
- Charitable donations.
- Professional consulting fees, legal, tax prep, information technology (IT).
- Business meals and travel.

These categories are but a sampling of the costs of running an electrical contracting company or any other business. The costs incurred from home office overhead expenses are shared among all jobs under contract. More exactly, a percentage of the total home office overhead is expected to be absorbed by each project over its life. These actual costs are tracked as expenses are incurred, weekly or biweekly, such as payroll, and monthly, such as rent or mortgage payments. Senior management, with the aid of the bookkeeping or accounting professional, make adjustments as dictated by the particular market conditions. This is why, when the economy declines, home office overhead costs are the first costs to be downsized. Adjustments can include the outright elimination of the cost such as the layoff of nonessential staff.

As noted previously in this chapter, home office overhead for a company that is well established and has created a market niche may vary only slightly year to year. The company has stabilized its indirect overhead expenses. Minor increases for salary hikes and new expenditures can often be offset by an increase in work under contract.

For new, growing companies, home office overhead can change weekly or monthly. Managing the indirect overhead can pose a real challenge to cash flow and in some circumstances paralyze the company if not handled correctly. Mismanagement of the indirect overhead costs is perhaps one of the main reasons why so many contractors are unable to realize a profit or even stay in business. This effect is manifested in two ways. Either a company does not know the true overhead cost and therefore fails to mark up its costs enough to recover them, or management does not restrain or control overhead costs effectively and fails to remain competitive.

If a contractor does not know the costs of operating the business, then these costs will probably not be recovered. Many companies survive, and even turn a profit, by simply adding an arbitrary percentage for overhead to each job, without knowing how the percentage is derived or what is included. This is the business version of Russian Roulette, and it is only a matter of time before failure.

In theory, the calculation of the overhead costs can be rather routine. It is the month-by-month collection and analysis of the overhead outlays as a function of the work in progress. While simple in theory, it is often a moving target and difficult to pin down. Accurate approximation is often sufficient and the best that can be hoped for. It is rarely an exact number. The following example is hypothetical and is meant as an illustration of the calculation of indirect overhead.

Example

For the fiscal period starting January 1 and ending December 31 of that year Sparky's Electric, Inc. calculated (using acceptable professional standards) that the indirect overhead for the company during that period was $345,500. It was also determined that the volume of work under contract in that same period that was completed was $2,891,000. Then the indirect overhead could be calculated as:

$$\text{Indirect overhead} = \$345,500 \div \$2,891,000 = 0.1195 \approx 12\%$$

Briefly, this example illustrates that it cost $345,500 in indirect overhead to execute $2,891,000 in work and that the home office overhead for the period (year) was 12%. For all projects that are bid in that period, the 12% was added to the cost to cover the proportional share of the overhead cost that the project will be expected to absorb if the bid is won and the work is performed.

If the annual volume of work under contract increased significantly (i.e., more than 7% to 10%), then the previously used percentage for overhead is no longer valid. If the volume increases without a proportional increase in indirect overhead expense, then the calculated 12% would decrease as follows:

$$\text{Indirect overhead} = \$345,500 \div \$3,380,000 = 0.102 \approx 10.2\%$$

The 10.2% indirect overhead would translate to a savings of overhead of 1.8% (12% − 10.2% = 1.8%) that would be moved to the profit column. If the volume of work decreased significantly, again without any change to the overhead expense incurred, then the 12% would increase as follows:

$$\text{Indirect overhead} = \$345,500 \div \$2,352,000 = 0.147 \approx 14.7\%$$

Conversely, this is interpreted as an increase of indirect overhead of 2.7% (14.7% − 12% = 2.7%). With a stipulated sum (fixed price) contract, that would mean that 2.7% would be moved from the profit column to the cost column to cover the increased overhead.

While an increase of 2.7% in overhead may seem small, it is 2.7% that is not being recovered by the project but must be deducted from the anticipated profit. One can easily imagine how this can go very bad, very quickly. One remedy that can be instituted quickly to reduce overhead is to reduce office payroll. Payroll is often the largest and most frequent cost associated with home office overhead. Unfortunately, this impacts a person's life and is one of the least pleasant tasks associated with owning a business.

The second point made is the lack of restraint or self-control of upper management (or the owner) resulting in a constant creeping increase in indirect overhead expense. Most professionals can immediately call to mind the meteoric rise of a company only to disappear from the scene shortly thereafter. Every marketplace is littered with them. Or the small company that is more concerned with image than substance and purchases new trucks, one after another, only to find that the overhead has made the company top heavy and cash poor. This problem is a result of business immaturity of the management and is not part of the scope of this book.

PROFIT AND CONTINGENCIES

Two crucial but difficult-to-calculate costs must be added to the estimate in the *recapitulation*, or Estimate Summary, prior to submitting the bid: profit and contingency. These topics are rarely addressed in most estimating texts on the market today. The most reasonable explanation for this lack of information is that determining the appropriate profit and/or contingency is a process that relies more on experience or judgment rather than calculating a quantity and pricing it, as is the case with wire. If 10 different estimators were queried, they would most likely say that profit is assigned as a percentage of the cost of the work. However, the actual decision-making process that led to that particular percentage would be different from one estimator to the next. The same applies to contingencies.

This section will introduce specific considerations to review before determining an appropriate profit and contingency (if even applicable) on a project-by-project basis. Although profit is the last number to be added to an estimate (with the exception of a performance and payment bond premium), the profit will be discussed before contingency, since not all projects warrant a contingency.

Profit

One of the main yardsticks for measuring the success of a construction project is profit. Without it, the project would be considered a failure. Profit can be loosely defined as the amount of money left after all of the bills have been paid. It is a cost to the client who pays the bills, but it is the only category in the contractor's estimate that does not have a commensurate cost that is incurred. It is the necessary component to make a business viable, fiscally healthy, and able to grow. Profit is the basis for our business economy and the capitalist system and is what allows a contractor to grow and hire more people. Predicting the correct amount of profit that a project can support is one of the most difficult tasks for the construction professional. Too small a profit, and the return does not warrant the risk taken. Too large a profit, and the bid can be lost to greed. Ideally, the amount of profit to be added should be the maximum the project can support, but just slightly less than the next bidder's.

It is generally acknowledged among construction estimating professionals that the cost of materials, labor, and equipment calculated by the electrical contractor's estimator will be roughly the same for most contractors bidding the same project. Some items will be higher, and some will be lower. However, in the end, all *costs* should be about equal. This also applies to subcontractors.

On bid day, subcontractors submit quotes to the majority of prime contractors bidding a project. Some subcontractors may have been solicited by a particular prime contractor, while others just "cover all the bases" by submitting bids to all of the bidders. If the statement that "cost is cost" is true, then it could be inferred that adding overhead and profit to the estimate can be the deciding factor in winning or losing a bid. Remember that most contractors compete; that is, there are other contractors that share their market. This means that for all intents and purposes the competitor's costs, direct and indirect, are similar. This is what allows them to share the market.

In some companies, determining profit is the responsibility of the estimator, while other contractors consider it to be the domain of senior management or company principals only. Irrespective of the party assigned this duty, it is clear that by the end of the estimate preparation, the person with the best understanding of the risk involved and the uniqueness of the project is the estimator. As a result, the estimator is the most likely candidate to contribute to the decision-making process in determining the profit.

Many texts use 10% for profit in their examples as the appropriate percentage to be added, regardless of the project. The author suggests that this is more for ease in the mathematics in the example than an industry standard. While acknowledging that 10% may be acceptable for some projects and some contractors, it is clearly not a one-size-fits-all process.

Factors in Determining Profit

The actual mechanical process of assigning profit to a project can be done two different ways: as a percentage of the cost of the work or as a fixed fee based on time.

Regardless of the selected method, many factors affect the determination of profit. Some are tangible and some less so. As with all portions of the estimate, careful consideration should be given to the reasons behind each decision. While acknowledging that there are no clear answers or step-by-step procedures for arriving at the correct profit for a project, there are a series of considerations that should be reviewed when determining the appropriate amount to apply. The following sections show the thought process that must take place before the bid is finalized and should provide guidance for properly assigning the right amount of profit. These guidelines are not presented in any specific order; their order of importance will vary depending on the individual project.

Risk versus Reward

All construction projects entail a certain degree of risk, which can manifest itself in many forms. As a means of offsetting the risk, specific management techniques are used to "share" it. For example, a prime contractor might secure a performance and payment bond from a subcontractor who has a large share of the work in order to assign some of the risk to another party. However, from a business standpoint, a project with a high degree of risk requires more management time, resources, and generally creates more of a strain on a company's infrastructure. As a result, the company should be compensated for the risk endured. In other words, risk must be rewarded, which, in the construction business, is defined as profit. Profit is the reward for risk assumed, managed, and triumphed over. The reward should match the risk, supported by the general theory that the more risk involved, the higher the profit should be. There is no magic formula to calculate profit as a function of risk, yet almost all reasoning for applying a specific profit to a project can be traced back to the risk involved. One must carefully evaluate the risk that the project will impose on the company and define or quantify it in terms that can be used to determine a profit.

Reputation in the Marketplace

All completed projects will have an impact on a company's standing in the marketplace. Many larger electrical contracting firms with marketing departments actively pursue projects that will enhance the chances of future work or that have high visibility in their sphere of influence. While it does not take a marketing genius to figure out that a high-profile project will receive more attention, this may not always be a good thing. Projects that are "built in the newspaper" or under the watchful eye of the public can be a management nightmare. Along with the normal management team, damage control specialists may be needed to help keep public opinion and rumors in check, as these types of projects have a momentum and dynamic all their own. They can have a tremendous impact on a firm's reputation and on future business opportunities. Be sure to consider what the successful completion of the project will mean for the company and its reputation in the marketplace. Conversely, it is always wise to also speculate on how a failure would affect the firm's reputation.

Schedule Impacts

A project's schedule greatly affects the amount of profit that should be added to the estimate. Projects with durations in excess of a year will affect the company's balance sheet for multiple years and will need to carry enough profit for the firm beyond the current year. Losses will affect more than the current year's balance sheet as well. It is a recognized fact that projects extending beyond one year in duration are more difficult to manage because of potential changes in the marketplace that cannot be accurately predicted at bid time. Wage increases, inflated material costs, availability of resources, and the economy in general are some of the variable factors.

Carefully review project durations that appear to be too short. Those with unrealistic schedules often require infusions of capital and extra management to be completed on time. While these costs can be accounted for in the estimate, you also need to consider the fact that you may not be able to perform other work at the same time, which means lost business opportunities.

By the time the estimate has been completed and you are ready to add profit, a construction schedule should have been developed—and refined. This is necessary to determine project duration for time-sensitive costs. The schedule should enable the estimator to support or reject the owner's timeline under the contract provisions.

Contract Documents and Team Relationships

The level of design development in the bid documents also has an impact on profit. The more complete the design, the less risk to the contractor. While the level of design development affects the amount of justifiable profit, it also may necessitate adding a contingency. (This will be addressed from another perspective later in this chapter under Contingencies.) A contractor's prior history and working relationship with the architects and engineers for the project is also critical. Successful relationships with architects, engineers, and even owners play a significant role in assigning profit. A contractor's being viewed as part of the "team," rather than as an adversary, will have a direct impact on the profit line. The contractor's expertise is seen as critical to a successful project and must be rewarded by allowing reasonable profit. Contractors involved with architects, engineers, or owners who have a reputation for taking a "hard-line" approach often add greater profit margins to their estimates to compensate for these adversarial relationships.

Contract Clauses

Many contractors/estimators interpret the tone of the contract (included in the project manual) as a precursor to the way the project will be administered. Are the general and supplemental conditions peppered with unfavorable contract clauses or punitive language toward the contractor? Does the contract have liquidated damages or penalties of any kind? If so, are they reasonable? Is there exculpatory language that absolves the owner and architect from responsibility for delay to the project? While the owner/general contractor (GC) will often downplay the use of penalties clauses, they are there for a reason. Should the relationship deteriorate, the owner/GC has the right to exercise his or her contractual rights, and in most cases they will. Review the contract clauses carefully, and if necessary, seek legal advice on specific language that may be a concern. If the final decision results in bidding the project, make sure that adequate profit is included to compensate for the risk.

Impact on the Company's Resources

When determining profit, be sure to address the following questions as they pertain to your company's needs and resources:

- Does the company have access to capable subcontractors, vendors, and suppliers to perform the work? Are a majority of vendors being carried in the bid unknown and untested?
- Does the company have enough of its own labor resources to self-perform work or augment underachieving subcontractors should it become necessary?
- Does the management staff have the skill sets necessary to administer and manage the project?
- Will the firm have to hire new individuals to supervise or manage the project? If so, would this be considered additional risk as a result of the unknown factors involved?

- Does the company have the working capital to finance the work between owner/GC payments? Not having adequate finances to capitalize a project puts tremendous tension on relationships with subs, vendors, and suppliers who are key to a successful project.
- If the project can be administered by the current personnel and infrastructure, what effect will it have on company morale?
- Will the project tax the company to the extent that no other projects will get their fair share of attention or management, or worse—that the company will not be able to handle other projects?
- All of these are necessary considerations to determine the impact on the company's resources, a key factor in assigning profit.

Repeat Business

Many estimators consider the potential for repeat business when applying profit. This is a very real and important consideration. Estimators and management teams often reduce their profit in the hopes that the owner will reward this behavior with repeat business. This is a common and sound business practice for many sectors of the construction industry. It is the hallmark of retail construction. Bear in mind, however, that too small a profit may not make future projects with a particular owner/GC attractive. It may also set a precedent for future contracts. For the contractor who does frequent business with a client, remember that "a contractor is only as good as his last job." Reducing profit in hopes of repeat business can often have a negative effect because projects with insufficient or marginal profit are frequently relegated to the "back burner" in favor of more lucrative ones. This can end the repeat business cycle that was being anticipated.

Project Location

Many desirable projects may be outside the company's normal sphere of influence. While this does not expressly mean one should not bid the project, it must be acknowledged that there are inherent problems that come with working outside one's known business area. These include travel time and related costs, subcontractors' and suppliers' abilities to service the location, and general unknowns of doing business with new building departments and inspectional services, as well as public utilities.

Bidding Strategies

Many estimators employ a bidding strategy for winning work, which encompasses a wide range of techniques meant to provide an advantage, such as tracking the workload of the competition to determine potential threats. Most contractors strive to create a market niche for themselves. The theory is that as you do repetitive work, the learning curve disappears, and, as a result, the firm becomes more financially successful. In doing so, the company fits into a niche that is shared by competitors. Contractors who competitively bid projects will find themselves frequently competing against the same firms. The ability to track who is busy and who is "hungry" is helpful, since the hungry contractor is more of a threat than the busy one. Other bidding strategies include unique means and methods, such as prefabrication or assembly off-site, which often helps the bidder be more competitive. A successful bidding strategy provides an edge over the competition. The assignment of profit is a very big part of the bidding strategy.

Specialization

One-of-a-kind projects with no comparison model warrant an increased profit. While there are very few projects in the residential/light commercial sector of the construction industry that have never been done before, unique projects often involve highly specialized contractors, thereby limiting competition. The fewer the competitors, the larger the profit that should be expected.

Current Workload

Frequently, the profit margin is determined by how much work a contractor has under contract. Contractors with sufficient work add larger profit margins, using the logic that if more work is going to be added, it will have to be highly profitable as it taxes the company's infrastructure. Contractors with minimal work under contract are prone to taking projects with little or no profit, as any cash flow is preferable to a negative profit-and-loss statement. While this is true, accepting low-profit work can be an extremely dangerous practice and is not advocated in any situation other than the most dire of circumstances.

Demographics

By virtue of their locations (and the requirements of the market), certain projects warrant a larger profit margin. For example, assuming that all other (construction) costs are equal, a residential project in an area with higher real estate values will typically be assigned a larger profit percentage than a similar residence built in a lower-priced area. Projects far from the beaten path may also warrant a higher profit margin just for the added inconvenience and difficulty presented.

Contingencies

Contingencies, or adding money to address an unknown condition, are the most misunderstood line items in an estimate. The estimator should try to anticipate any costs that are not capable of being recovered if discovered. There are two schools of thought on contingencies:

Always Add a Contingency

Contingencies should be added for costs that cannot otherwise be recovered. Justifications for contingencies include call-backs that are not the contractor's responsibility but are sometimes done to further a firm's reputation. Consider the residential contractor who repairs damage caused by an unidentifiable party. Repairing the damage is a good business move and keeps the client happy. It also portrays the contractor as a reputable businessperson who stands behind his/her work even when there is a question of who is responsible.

Other scenarios include adding contingencies for weak or incomplete documents. It is not uncommon for architectural and engineering services to be kept to a minimum in the design stage of a simple project. The owner's logic is often that if an engineer's services can be kept to designing only the essentials, a good electrical contractor can flush out the details and make the design work. Even the most conscientious estimator cannot anticipate every condition, unforeseen or otherwise. Again, adding a contingency to help the client over some unexpected costs goes a long way toward future business. However, it is important to know when enough is enough.

For owners or clients developing a budget for financing or appropriations is wise to consider a contingency. This covers the unknowns in renovation or even new work. The contingency should be for a specific reason, for example, escalation between the time a budget is developed and the project comes to fruition.

Never Add a Contingency

Some people feel that adding any money to an estimate that is not applied to a tangible, defined cost or scope of work is a sign of poor estimating practice. It is the purpose of an estimate to accurately anticipate all costs to be incurred in a project. The contract documents, plans, and specifications act as the basis for the estimate. The drawings represent the quantity, and the specifications represent the quality. If an item or scope of work in question is not shown on the drawings or called out in the specifications, it is outside of the scope of work and extra to the contract. Adding money to the estimate for work that is not defined at bid time is often considered irresponsible and can dramatically affect hard competitive bids. Most professionally drafted contracts provide relief for added scope.

Deciding if Contingencies Are Necessary

Other more general questions arise as a result of the adding contingencies. For example:

- What if the amount of the contingency is insufficient and does not cover the cost of the work?
- What if the amount is too much and reduces the competitiveness of the bid?
- Does performing work at no cost to the owner under the guise of a contingency create a dangerous precedent for future uncovered problems?
- Should the owner be aware of the contingency and its amount?
- If the contingency is not spent, is it returned to the owner?

The answer to the last question can often shed light on the decision.

It is clear from the discussion in this chapter that there are no hard-and-fast rules for assigning profit or contingencies to an estimate, but merely considerations that must be carefully reviewed for each and every project. Ample thought beforehand, paired with varied experience, will help estimators arrive at the appropriate profit margin for the individual company and project.

8 The Unit Price, Project Overhead Summary, and Estimate Summary Sheets

This chapter will introduce the reader to the more common sheets in the pricing phase of the estimating process. There are three common sheets that estimators recognize regardless of whether the estimating is done by hand or by software. The first is called the *unit price* sheet or cost sheet. This is where individual quantities of a task are multiplied by the various unit prices and extended to a subtotal. The next in the trio is the *Project Overhead Summary* sheet. This sheet tabulates the cost of the direct or project overhead of the General Requirements. The last is called the *Estimate Summary* sheet and is the recapitulation of the estimate.

Each of these plays a key role in estimating costs and the compilation of those costs in arriving at a final bid number. In the following sections the author will explain each sheet, its use, and its relationship to the others and to the process as a whole.

While computerized estimating software creates a variation of each of the three sheets, and performs the appropriate calculations, it is important for the estimator to understand the process.

UNIT PRICE SHEET

The unit price sheet is the most elemental level of the pricing process. It is set up in a columnar format that will be used as a spreadsheet for calculating the cost of a specific category of tasks. Categories are separated by CSI MasterFormat numbers for ease of identification. It illustrates each task with a concise but complete description.

This is often a direct transfer from the quantity takeoff of the task. The task is quantified in a standard or industry defined unit of measure. Each successive pair of columns has a column with the unit price column and the extended total column for material, labor and equipment. Table 8.1 illustrates a typical unit price sheet created using Microsoft Excel.

Consider task 1.01, the first task estimated in Table 8.1. This line is identified under the CSI MasterFormat section to which it belongs. Next is the description in words and numbers identifying the task; *Incandescent, high hat can, round alzek reflector, 100W*. It is best to provide a description that includes sufficient information to identify the task. To the right of the description is the total quantity *(56)* for that task, followed by the unit of measure *(EA)*.

Adjacent to the unit of measure column are two columns under the heading for Materials, Labor, and Equipment. The first column in each is for the *unit price* for task 1.01. The second column in each is called the *extension*. It is the product of multiplying the total quantity by the unit price, for example, line 1.01 for the material: *56 EA × $70.00 = $3,920.00*. This number represents the total material cost for task 1.01. The same calculation is done for the Labor and Equipment columns.

Each of the extensions is then added to arrive at the total cost for line 1.01. Each column is totaled down and to the right. The lower right hand corner total represents the unburdened cost for that category. To the right of that is the Check column. This acts as an arithmetical check and balance of the total.

There are many variations of the unit price sheet. One of the more common versions includes a third column that calculates labor-hours as well. Table 8.2 illustrates the added labor-hours column.

The last line representing the total cost for each CSI MasterFormat section will be transferred to the Estimate Summary sheet. This summary line is exclusive of sales tax, home office overhead, profit, and bond cost. This will be added in the final stages of recapitulation.

PROJECT OVERHEAD SUMMARY SHEET

As discussed in Chapter 6, "Direct Costs," each project has distinct costs that must be added to cover the direct or project overhead required for each project. These costs, while often of a nonproduction nature, are no less important to the accurate estimating of a project. They include a wide variety of costs such as temporary facilities, trailer rental, supervision, telephone usage, dumpsters, and electrical power usage. All are costs that are directly related to one specific project, and only one project.

These costs are categorized as either fixed costs and time-sensitive costs (see Chapter 6, "Direct Costs"). Since many of them are common to all projects such as supervision or permits, estimators frequently employ a checklist to ensure that all requirements are recognized. Table 8.3 is a typical Project Overhead Summary sheet

Table 8.1 Typical Unit Price Sheet Using Microsoft Excel

Bid Date:	July 30, 2014	Project:	Renovated Office Building						Estimate for:	Base Bid
Time:	2:00 PM		10 Washington St., Boston, MA						Addenda:	#1, #2 have been acknowledged
Estimator:	Jack Smith	Project No.	2014-23A						Checked By:	Mike Jones

SECTION	DESCRIPTION	QUANTITY	UNIT	MATERIAL		LABOR		EQUIPMENT		TOTAL	Check
				UNIT COST	TOTAL	UNIT COST	TOTAL	UNIT COST	TOTAL	TOTAL	
26 51 00	**Interior lighting**										
1.01	Incandescent, high hat can, round alzek reflector, 100W	56	EA	$70.00	3,920.00	$52.50	2,940.00	$0.00	0.00	$6,860.00	
1.02	Metal halide, 2' W × 2' L, 250W, integral ballast, recess	33	EA	$334.55	11,040.15	$152.50	5,032.50	$12.30	405.90	$16,478.55	
1.03	HP sodium, 2' W × 2' L, 150W, integral ballast sur. mtd.	12	EA	$463.20	5,558.40	$155.50	1,866.00	$14.50	174.00	$7,598.40	
1.04	Wall-mounted, metal cylinder, 75W, painted finish	16	EA	$55.00	880.00	$43.00	688.00	$0.00	0.00	$1,568.00	
1.05	Vaportight, incandescent, ceiling mounted, 200W	23	EA	$81.30	1,869.90	$67.55	1,553.65	$0.00	0.00	$3,423.55	
1.06	Low bay aluminum reflect, 250W	33	EA	$363.00	11,979.00	$131.25	4,331.25	$0.00	0.00	$16,310.25	
1.07	2' x 4' fluor., drop in, acrylic lens, 4-32W, T8 lamp	66	SF	$66.50	4,389.00	$88.40	5,834.40	$0.00	0.00	$10,223.40	
	Interior Lighting Totals				$ 39,636.45		$ 22,245.80		$ 579.90	$ 62,462.15	62462.15

Table 8.2   Price Sheet with Added Labor-Hours Column

Bid Date:	July 30, 2014	Project:	Renovated Office Building			Estimate for:	Base Bid
Time:	2:00 PM		10 Washington St., Boston, MA			Addenda:	#1, #2 have been acknowledged
Estimator:	Jack Smith	Project No.	2014-23A			Checked By:	Mike Jones

| | | | | MATERIAL | | LABOR | | | EQUIPMENT | | | |
SECTION	DESCRIPTION	QUANTITY	UNIT	UNIT COST	TOTAL	UNIT COST	TOTAL	Labor Hours	UNIT COST	TOTAL	TOTAL	Check
26 51 00	**Interior lighting**											
1.01	Incandescent, high hat can, round alzek reflector, 100W	56	EA	$70.00	3,920.00	$52.50	2,940.00	49.73	$0.00	0.00	$6,860.00	
1.02	Metal halide, 2' W × 2' L, 250W, integral ballast, recess	33	EA	$334.55	11,040.15	$152.50	5,032.50	85.12	$12.30	405.90	$16,478.55	
1.03	HP sodium, 2' W × 2' L, 150W, integral ballast sur. mtd.	12	EA	$463.20	5,558.40	$155.50	1,866.00	31.56	$14.50	174.00	$7,598.40	
1.04	Wall-mounted, metal cylinder, 75W, painted finish	16	EA	$55.00	880.00	$43.00	688.00	11.64	$0.00	0.00	$1,568.00	
1.05	Vaportight, incandescent, ceiling mounted, 200W	23	EA	$81.30	1,869.90	$67.55	1,553.65	26.28	$0.00	0.00	$3,423.55	
1.06	Low bay aluminum reflect, 250W	33	EA	$363.00	11,979.00	$131.25	4,331.25	73.26	$0.00	0.00	$16,310.25	
1.07	2' × 4' fluor., drop in, acrylic lens, 4-32W, T8 lamp	66	SF	$66.50	4,389.00	$88.40	5,834.40	98.69	$0.00	0.00	$10,223.40	
	Interior Lighting Totals				$ 39,636.45		$ 22,245.80	376.28		$ 579.90	$ 62,462.15	62462.15

created using Microsoft Excel. Although the General Requirements (project overhead) are Division 1, they are estimated last. Since many of the costs in project overhead are experience based and subjective, it requires a full knowledge of the project to accurately estimate these costs. The theory is that by the time the estimate cost for Divisions 2 through 46 is complete, the estimator will have an intimate understanding of the project and be in the best position to apply project overhead costs.

The Project Overhead Summary sheet follows a similar format to the unit price sheet, although many of the task's units of measure are in time such as days, weeks, or months. Similar to the unit price sheet, when complete the totals on the last page are transferred to the Estimate Summary sheet for inclusion in the total.

THE ESTIMATE SUMMARY SHEET

This is the point in estimating process in which much of the hard work, the takeoff and pricing, has been completed. Now all of the pricing has to be compiled, markups added, and summarized to arrive at the final total price. While the process is relatively straightforward, it should not be trivialized. Errors in this phase result in an erroneous total that leads to winning the award due to omitted costs.

The Estimate Summary is officially known as the *Recapitulation Sheet*. It is the compilation of all estimated categories of work that comprise the project, organized in rows by CSI MasterFormat, and divided into columns labeled *Materials, Labor, Equipment, Subcontractors,* and *Totals*. Each row is totaled across and each column is totaled down. This provides a check of the arithmetic in the lower right-hand corner. Table 8.4 is an example of an Estimate Summary Sheet created using Microsoft Excel.

The estimator transfers the materials, labor, and equipment costs from the individual estimating sheets to the summary sheet. Subcontractor costs are transferred from the individual quote after careful review and determination that the price is comprehensive and represents that desired scope.

Material quotes can also be included in the Estimate Summary in the materials column for a particular category of the estimate. Finally, the far right column typically has space for comments or remarks that relate or qualify that specific line in the estimate.

In day-to-day practice, many quotes from vendors and subcontractors are not submitted until the last minutes before the bidding deadline. The reasons for this are many and varied but often stem from an attempt to thwart bid shopping. Therefore, a system is needed to efficiently and quickly allow the recalculation of the entire estimate as a new price is added or substituted. This is where the spreadsheet applications of most estimating software programs excel. Last-minute quotes can be entered and the entire estimate total can be recalculated in a matter of seconds with a few keystrokes.

Table 8.3 Project Overhead Sheet

Bid Date:	July 30, 2014		Project:	Renovated Office Building				Estimate for:		Base Bid	
Time:	2:00 PM			10 Washington St., Boston, MA				**Addenda:**		#1, #2 have been acknowledged	
Estimator:	Jack Smith		**Project No.:**	2014-23A				**Project Duration:**		11 months	
Checked by:	Mike Jones										

01 00 00	Description	Quantity	Unit	MATERIAL		LABOR		EQUIPMENT/SUB		TOTAL	Remarks
				Unit Price	Extension	Unit Price	Extension	Unit Price	Extension		
1.01	Office trailer	11	MONS	390.00	4,290.00	200.00	2,200.00	—	—	6,490.00	
	a. Furnishings	1	LS	200.00	200.00		—		—	200.00	
	b. Setup and delivery	1	EA		—			240.00	240.00	240.00	
	c. Breakdown and return	1	LS		—		—	240.00	240.00	240.00	
	d. Trailer cleaning	11	MONS		—	110.00	1,210.00		—	1,210.00	
	e. Fax machine										
	f. Computer/Printer										
	g. Office supplies	11	MONS	80.00	880.00				—	880.00	
	h. Software										
	i. Ramps to trailers	1	LS		—		—	180.00	180.00	180.00	
1.02	40' Storage trailers	6	MONS		—		—	300.00	1,800.00	1,800.00	
	a. Delivery and pickup	1	LS		—		—	100.00	100.00	100.00	
1.03	20' Storage trailers										
	a. Delivery and pickup										
1.04	Telephone service—Landline										Use cell phones
	a. Install and removal										
	b. Monthly service per line										
1.05	Temporary electric										

01 00 00	Description	Quantity	Unit	MATERIAL		LABOR		EQUIPMENT/SUB		TOTAL	Remarks
				Unit Price	Extension	Unit Price	Extension	Unit Price	Extension		
	a. Office trailers hookup	1	EA		—	340.00	340.00		—	340.00	
	b. Trailer consumption	11	MONS	180.00	1,980.00		—		—	1,980.00	
	c. Project consumption										By owner
1.06	Water cooler consumption	11	MONS	45.00	495.00		—		—	495.00	
1.07	Thermometer	1	EA	25.00	25.00		—		—	25.00	
1.08	Temporary toilets	22	MONS		—		—	114.00	2,508.00	2,508.00	2 EA men's and women's
1.09	Temp. construction fence										
1.10	Staging										Use Personnel Lifts
	a. Set up/dismantle										
	b. Monthly rental										
1.11	Personnel Lifts										
	a. Delivery and pickup	2	EA		—		—	100.00	200.00	200.00	
	b. Monthly rental	5	MONS		—		—	340.00	1,700.00	1,700.00	
1.12	Small tools and equipment	11	MONS		—		—	200.00	2,200.00	2,200.00	
1.13	Temporary water										
	a. Hook up/dismantle										
	b. Consumption										
	c. Fees										
1.14	Temp. heat (if not electric)										By electric
	a. Trailers										
	b. Project										

(Continued)

61

Table 8.3 (*Continued*)

01 00 00	Description	Quantity	Unit	MATERIAL Unit Price	MATERIAL Extension	LABOR Unit Price	LABOR Extension	EQUIPMENT/SUB Unit Price	EQUIPMENT/SUB Extension	TOTAL	Remarks
1.15	Temporary protection										Not required
1.16	Winter protection										Not required
	a. Plowing										
	b. Enclosures										
1.17	Fork lift or lull										
1.18	Crane										
1.19	Project photos										
1.20	Tree protection										
1.21	General cleaning—ongoing										
1.22	Final cleaning										
1.23	Materials handling & distribution	40	HOURS		—	85.00	3,400.00			3,400.00	
1.24	Project sign	1	EA	600.00	600.00	300.00	300.00			900.00	
1.25	First aid kits	1	EA	110.00	110.00		—			110.00	
1.26	Temporary fire protection	1	LS				—	340.00	340.00	340.00	
1.27	Dumpsters—30 CY	6	EA	650.00	3,900.00		—		—	3,900.00	
	Dumpsters—recycling										
1.28	Pest control										
1.29	Cutting and patching										
	a. Coring bits	6	EA	350.00	2,100.00		—		—	2,100.00	
	b. Purchase coring machine/bits										
1.30	Permits										
	a. Electrical permit	1	LS	3,495.00	3,495.00		—		—	3,495.00	
	b. Occupancy permit										
	c. Misc. fees										

01 00 00	Description	Quantity	Unit	MATERIAL		LABOR		EQUIPMENT/SUB		TOTAL	Remarks
				Unit Price	Extension	Unit Price	Extension	Unit Price	Extension		
1.31	Police details/flagman										
1.32	Layout										
	a. Registered										
	b. Own forces										
1.33	Testing										
	a. Soil testing										
	b. Concrete testing										
	c. Load testing										
1.34	Misc. hardware	11	MONS	100.00	1,100.00		—		—	1,100.00	
1.35	Pickup trucks										
	a. Gasoline usage	45	WKS	70.00	3,150.00		—		—	3,150.00	
1.36	CPM schedule—initial devel.										
	a. Update CPM Monthly										
1.37	Dewatering										
	a. Localized dewatering										
1.38	Special safety equipment										
1.39	Attorney's fees										
1.40	Interior barricades										
1.41	Record drawings				—		—				
	a. CAD fee for record docs	1	LS		—		—	550.00	550.00	550.00	
	b. Printing/Reproduction										
	c. Mylars				—				—	—	
1.42	Project closeout	2	PHASES	450.00	900.00	800.00	1,600.00		—	2,500.00	

(Continued)

Table 8.3 (*Continued*)

01 00 00	Description	Quantity	Unit	MATERIAL		LABOR		EQUIPMENT/SUB		TOTAL	Remarks
				Unit Price	Extension	Unit Price	Extension	Unit Price	Extension		
1.43	Site security										
	a. Watchman										
	b. Custodial overtime										
1.44	Utility company charges										
	a. Electric										
	b. Cable/Tel.										
1.45	Special requirements										
1.46	Insurance										
	a. Builder's risk	1	LS	557.00	557.00		—		—	557.00	
	b. 3-yr extended comp operations										
	c. 10M add umbrella										
1.47	Punchlist										
1.48	Personnel										
	a. Superintendent/ Foreman	47	WKS			2,355.00	110,685.00		—	110,685.00	
	b. Asst. superintendent/ Foreman										
	c. Project manager	22	WKS		—	2,400.00	52,800.00		—	52,800.00	
	d. Administrative staff										
	TOTAL				23,782.00		172,535.00		10,058.00	206,375.00	Ck 206,375

Table 8.4 **Estimate Summary Sheet**

Bid Date:	July 30, 2014	Project:	Renovated Office Building			Estimate for:	Base Bid
Time:	2:00 PM		10 Washington St., Boston, MA			Addenda:	#1, #2 have been acknowledged
Estimator:	Jack Smith	Project No.	2014-23A			Checked By:	Mike Jones
Sect.	Description	Materials	Labor	Equipment	Subcontractor	Totals	Remarks
01 00 00	Project Overhead Summary Sheet	23,782.00	172,535.00	10,058		206,375	See Project Overhead Summary Sheet
01 20 00	Allowances	16,500.00				16,500.00	Walkway bollards per specs—materials only del.
26 05 05	Electrical Demolition		6,785.00	2,356.00		9,141.00	
26 05 33	Raceways and Boxes for Electrical Sys.	12,567.00	19,345.00	1,789.00		33,701.00	
26 05 48	Vibration & Seismic Controls for Electrical					—	
26 09 23	Lighting Control Devices					—	
26 08 00	Commissioning of Electrical Systems				22,000.00	22,000.00	Johnson Electrical Commissioning, Inc.
26 22 00	Low-Voltage Tansformers	4,592.00	7,390.00			11,982.00	
26 24 00	Switchboards and Panel Boards	27,980.00	45,345.00	1,340.00		74,665.00	
26 27 16	Electrical Cabinets and Enclosures	6,324.00	5,567.00			11,891.00	
26 29 23	Variable-Frequency Motor Controllers	6,789.00	3,450.00			10,239.00	
26 41 00	Facility Lightning Protection				15,000.00	15,000.00	Sparky's Lightning Protection Co.
26 51 00	Interior Lighting	39,636.45	22,245.80	579.90		62,462.15	Mat.-Boston Lighting Co. quoted dated 7-28-14
26 53 00	Exit Signs		3,300.00			3,300.00	Materials included in 26 51 00

(Continued)

Table 8.4 (*Continued*)

Sect.	Description	Materials	Labor	Equipment	Subcontractor	Totals	Remarks
26 56 00	Exterior Lighting	15,301.00	5,690.00	3,457.00		24,448.00	Boston Lighting Co. 7-28-14 incl labor for bollards
27 30 00	Voice Communications				34,567.00	34,567.00	All Data, Inc quote 7-29-14
28 31 00	Fire Detection and Alarm				86,799.00	86,799.00	Acme Fire Alarm Co. quote dated 7-29-14
	SubTotal	153,471.45	291,652.80	19,579.90	158,366.00	623,070.15	$ 623,070.15
	Massachusetts Sales Tax	9,591.97	6.25%	1,223.74		10,815.71	
	Subtotal					633,885.86	
	Home office overhead	12.70%				80,503.50	
	Subtotal					714,389.36	
	Profit	8.00%				57,151.15	
	Subtotal					771,540.51	
	Performance and Payment Bonds	0.93%				7,175.33	
	Total					778,715.84	
	BID					$778,716.00	

Like all other segments of the estimating process there are guidelines that help to make the process efficient and reduce errors. Some general guidelines for this process are as follows:

- A single individual on the estimating team, preferably a senior or chief estimator, should control the transfer of data to the Estimate Summary Sheet.
- A single individual on the estimating team, preferably a senior or chief estimator, should control the edit of inputs on the Estimate Summary Sheet.
- Placeholders or *plugs* should be entered in open cells while waiting for quotes from subs or vendors. This allows the estimate to develop an order of magnitude to it.
- No subcontractor or vendor quotes should be added to the Estimate Summary Sheet until each has been carefully reviewed and qualified for scope and comprehensiveness.
- A single individual on the estimating team, preferably a senior or chief estimator, should control the Estimate Summary Sheet.
- Costs for all items should be rounded to a nearest whole number (no cents). This helps prevent confusion during the transfer of costs.
- The Remarks column should identify the subcontractor or vendor carried in the line, and the date or version of the quote carried, especially if there are multiple iterations of the quote.
- Project Overhead (direct overhead) costs for the project should have a separate line in the Estimate Summary.
- The Estimate Summary sheet should have clear subtotals for each column of cost.
- Allowances that are required to be carried in the base bid should be identified in a separate line and listed correctly by category: material, labor, equipment, and so on.
- Subcontractor quotes should be listed separately. These costs contain the subcontractor's markups and may be treated differently from other direct costs when the estimator calculates the prime contractor's overhead and profit.
- Alternates, whether add or deduct, should be done on a separate Estimate Summary sheet. This prevents confusion between the Base Bid and any Alternates.
- Whenever possible, the transfer of costs to the Estimate Summary sheet should be checked by another individual to avoid transposition errors.

Applying Markups at the Estimate Summary Level

As illustrated in Table 8.4, the numbers are totaled and checked to arrive at the unburdened or *raw* cost of the project. This is represented by the first subtotal below the double line. Raw cost can be defined as the totals of all of the materials, labor, labor burden, equipment, subcontractors, and project overhead. What it does not include is the sales or use taxes (if applicable), home office overhead, profit, and bond costs (if applicable). All of the modifiers or percentages added to the subtotals below the double line are commonly called the *markups* since they modify the cost, specifically, the sales tax, home office overhead, profit, and performance and payment bonds.

The author would like to remind the reader that as noted in Chapter 6, "Direct Costs," the transferred labor cost that results in the subtotal of the labor column includes wages, benefits, fixed taxes, and insurances, but does not include home office overhead, profit, and sales or use taxes (if applicable).

The reader should also be aware that markups are compounded in the Estimate Summary sheet below the double line. This means that the markup percentage for profit is on the subtotal just before it, which includes home office overhead. That results in charging profit on the costs incurred for home office overhead. This is the correct method, despite much debate to the contrary. The justification is simple—profit is the reward for risk that will be incurred on a project. Risk on a project comes in many forms. The cost of a project is considered a risk. The more a project costs, the more risk to the contractor. The absence of profit markup on the cost of home office overhead would be the equivalent of loaning the dollar value of the home office overhead without interest—not a wise business decision.

The markup for Performance and Payment bonds is the last item added since the surety calculates their premiums on all costs and profit, the contract total. The only exception would be the application of taxes on the entire contract.

Markups on Subcontractors

Many estimators do not believe in a one-size-fits-all for profit. That is, some estimators have bidding strategies that will apply different profit margins on different categories of work based on a varied perception of risk. For example, self-performed work might rate a higher percentage of profit than subcontracted work or vice versa. If one were to apply this theory to Table 8.4, it would require two different percentages for profit applied to the subtotals for subcontractors versus material, labor, and equipment. This is a perfectly acceptable practice if an estimator can define tangible differences in the risk. Considerations may include:

- Is the subcontractor providing a Performance and Payment bond to the prime (electrical) contractor?
- How much of the project is self-performed and how much is subcontracted?
- How often has your company done business with this subcontractor, and how successful has the relationship been?
- How financially stable is the subcontractor? Does he or she require more frequent payments than the terms of your contract with the owner?
- Is the subcontractor reliable and independent or do they require micromanagement?
- What is the project duration?

These are merely some of the considerations for determining the profit that should be applied to the subcontractor's price. There is no standard or acceptable range; it depends on what the market will bear. If the prime contractor adds too much, the bid will not be competitive. If the amount is too little, the bid could be too low, which would result in substandard return or worse—performing the work for free. Neither of these is the hallmark of a successful contractor.

9 | Prebid Scheduling

The value of planning and scheduling is evident once the contract is signed and work commences on the project. However, scheduling is also important during the bidding stage. Not to the extent that one would schedule for performing the work but for different reasons. The following are just a few of those reasons:

- To determine if the project can be completed in the available time using a normal workweek schedule
- To determine if normal crew sizes will complete the work within the available time
- To identify potential premium time requirements
- To determine the time requirements for supervision
- To anticipate possible temporary costs for facilities
- To price certain General Requirement items and project overhead costs
- To identify weather considerations that may impact costs
- To budget for equipment usage
- To coordinate and evaluate delivery dates for long lead items
- To help evaluate supplier quotes relative to delivery times

The aforementioned is by no means a comprehensive list, but an introduction to the scheduling considerations that should come to mind during the bidding process. While the schedule produced prior to bidding is not as detailed as a progress schedule during the work, it still must address major categories of work, phases, milestones, lead times, and all constraints imposed by the contract.

This chapter introduces the reader to basic scheduling theory and the Critical Path Method (CPM) of scheduling in general. It is not meant to be a detailed explanation of CPM, but merely a discussion, focusing on how the schedule is used during bidding.

SCHEDULING THEORY

Scheduling is a decision making process that is logically complex but very simple mathematically. It is a step-by-step map through the project allotting time to perform each of the steps called *tasks* or *activities*. If estimating is for determining the cost of a task, then scheduling is for determining how long it will take to perform the task.

Construction schedules have some basic similarities, the most common of which is that work is measured as a function of time. Another is that the project is decomposed into incremental parts that represent an action required to achieve the deliverable. For example; *Set 4" steel conduit in trench* or *Install 2'×4' drop-in fluorescent fixture in grid.*

Tasks are performed in a specific order and often have an interdependency between each other. Tasks can occur sequentially or simultaneously. There is also more complicated sequencing of tasks. The sequencing of tasks is referred to as the *relationship* between tasks. Tasks in a schedule must be adequately described so that a non-involved party can understand what is occurring in the task. Most tasks should describe some type of action.

A schedule must have *milestones* or mini goals for measuring incremental progress. Tasks should also have a *resource* assigned to perform each task. Resources can be a single individual such as an electrician or it can be multiple tradespersons. Ideally, though not always the case, resources should be applied efficiently in the schedule. Tasks consume time to perform, and so each task must have a duration, adequate to accomplish the task. The duration of the task is measured in *workdays*, although the overall schedule is measured in *calendar days.* Workdays are most often Monday through Friday, and a single workday is 8 hours. Major holidays are typically non-workdays. In contrast, there are seven calendar days in a week and the week is measured from Sunday to Saturday. Workdays can be changed to include Saturdays and Sundays to reflect the specific needs of the project schedule.

Schedules must be logical; that is they must make sense both technically as well as chronologically. They usually follow the sequence dictated by normal construction practice, where physical relationships come first, followed by contractual relationships, and then managerial relationships.

In addition to the text portion which describes the task, and its duration, schedules employ a very powerful visual tool that defines the work with symbols and illustrates the relationships between tasks. Schedules should illustrate major milestones, contractual obligations or constraints, and above all a start and finish date for each task and the project as a whole. Schedules should also delineate time in calendar days and in months of the year.

CRITICAL PATH METHOD (CPM)

The *Critical Path Method* of scheduling, abbreviated *CPM* is uniquely fitted for the construction industry. In short, the concept behind the CPM schedule is that there are a series of sequential tasks that are linked from the start to the end of the project. If any one of these tasks is delayed, it will delay the finish date of the project as a whole. It is based on the fact that these tasks, called *critical tasks* are separate by inextricably linked. The start and finish of each critical task impacts the start and finish of the succeeding critical task. The critical path is the chain of critical tasks that must start and finish on time to keep the project as a whole on time. It is the longest path of planned activities to the finish of the project.

Not all tasks are critical, some tasks can occur at the same time as critical tasks, these are called *concurrent* tasks. Critical tasks can also be defined as those tasks in which the amount of time available to perform a task is equal to the time required to perform the task. For concurrent tasks, the time available is always greater than the time required to perform the task. This extra time is called *float* and is calculated in workdays.

When tasks are connected or *linked*, it illustrates the *interdependency* between them, the sum of their individual durations along the critical path determines the overall duration of the project. It thereby allows the estimator to determine when along the timeline the work will occur, as well as its relationship to key project milestones and seasons of the year.

By its very design, the CPM allows the estimator to illustrate the progress of the work as envisioned at bid time. The CPM also employs a visual format with the added component of interdependency.

While the overwhelming majority of industry practitioners schedule with some type of software it is essential to understand what goes on inside the CPM scheduling process. This text assumes the reader understands the mechanics behind CPM and will avoid any discussion of the forward and backward pass and the arithmetical computations associated with them.

VALUE OF THE PREBID SCHEDULE

As with almost all vertical construction today, the network scheduling method of choice is the CPM. It has a tremendous value in the initial planning process that often occurs during bidding, well in advance of the award. To accurately determine the duration of time-sensitive costs, a schedule must be drafted. Being most familiar with the project, the estimator is the likely and most qualified candidate to develop the schedule. This should be done for any estimate, any project—residential or commercial—regardless of size, in order to accurately predict time-sensitive costs. The prebid schedule is typically done when all of the takeoff and pricing is complete with the exception of Division 1—General Requirements.

One of most important questions answered by a prebid CPM schedule is if the time provided is sufficient to complete the work. It is not uncommon for the electrical contractor to be required to conform to someone else's schedule, namely that of the owner or GC. In preparing a prebid schedule, the electrical estimator can ascertain independently whether the work can be completed within the available time for performance with the available workforce. It can also provide a basis for determining a crew size or even the need for extended work hours that may be required to meet the schedule. It is highly recommended that the project schedule be determined independently of any provided schedule. The initial schedule should be developed with normal working times in mind. If the critical path needs to be compressed in order to achieve substantial completion on the required date, then the costs associated with the acceleration must be included as part of the estimate. The costs of acceleration that may be required to meet contract time for performance are sometimes overlooked.

For projects that will be built in climates that have very different seasonal temperatures or weather, a schedule is essential. It allows the estimator to determine if added costs for weather protection enclosure or temporary heating is required to perform a task based on the time of year in which it will be done. For labor-intensive tasks that occur during periods of weather extremes (hot or cold), the estimator can consider adjusting productivities for the various tasks that occur during each period.

Most professionals acknowledge that project overhead cost can be either fixed, such as an electrical permit or can vary with time such as a supervisory costs or the rental of equipment. Variable costs are schedule-driven and are determined based on the length of time the items or services are needed on the project. Variable costs can also be thought of as more subjective in the sense that they are frequently defined based on experience and judgment. Variable, or time-sensitive, costs often require that a "means and method" be adopted to develop the estimate. The schedule allows the estimator to determine the duration of specific variable overhead costs and when they will become unnecessary. Time-sensitive costs have a direct relationship to the project progress schedule and any project phasing plan.

Vendor delivery dates can be loaded onto the schedule and analyzed to determine if they will arrive when needed or if some expedited process may need to be considered. This simple analysis can be used in determining which vendor should be awarded the project based upon their ability to delivery on time.

Depending on the complexity of the project, several generations of schedules may be required as one becomes more familiar with the project or as input is added from other members of the project team such as subcontractors or vendors. It should be noted that the schedule used for estimating purposes is not the same version that might be required after the job is awarded. It is, however, a primer for the schedule that will be developed after award, as it provides insight as to the estimator's thought process at bid time.

A prebid schedule can provide much more information than simple job duration. It can be used to refine the estimate by introducing realistic labor projections. The schedule may also help the contractor to adjust the structure and size of the company based on projected requirements for months, even years, ahead. A schedule can also become an effective tool for managing contracts and communicating expectations.

TASK TYPES

Every estimator understands that there is far more to a schedule than just the tasks of installing the physical components of the work. There are permits to be issued, submittals to be approved, equipment to be ordered, and items to be fabricated off site, to name but a few. All of these items have a tremendous impact on the production work; the actual field tasks.

Consider the installation of a piece of switchgear on a project. Before the switchgear can be installed it has to be fabricated, and before the fabrication, there has to be detailed shop drawings drafted that show compliance with the specified product. These drawings are reviewed and approved by the electrical engineer, or corrections may be needed and a resubmittal may be required. In any event, it is evident that each step in the process is necessary and contingent upon the step or steps before it. Imagine if the shop drawing review took 4 weeks longer than scheduled. Would that delay the fabrication of the switchgear? You bet. If the fabrication is delayed, then one can understand how the delivery would be delayed, and then the installation of the switchgear delayed, and so forth all the way to the energizing of the project.

So if a task such as shop drawing review and approval can have that dramatic an impact on the schedule and the actual production work, should it be included in the schedule? The answer is—maybe! A brief note of caution is due here. Remember the prebid schedule is not the final performance schedule. It is used for determining key factors in the project relative to the estimate and the overall length of the project. This is where an ample helping of professional experience and judgment can make a difference.

TASK DURATIONS

Duration is defined as the time it takes to execute a task. It is the difference between a task's start date and its finish date. As previously noted, durations are measured in workdays.

Workdays are the appropriate increment because there is a direct correlation to productivity and the daily output of the individual or the crew. Productivity is measured by the daily output for a crew.

Determining the duration of a task is less a matter of experience than a mathematical calculation. That is not to say that experience does not play a part. There are two primary methods in which individual task durations are determined. Both require that a few general rules are followed:

1. Assume that each task will be performed normally. In this circumstance, the term *normal* defines ideal or near ideal working conditions: temperate climate, appropriate tools, adequate materials close by, sufficient working space, and so on. This ideal set of conditions translates to the most efficient productivity for the crew and as a result the lowest unit cost. It is not unrealistic to expect that a task will be performed at the most efficient productivity. In fact it is fairly common to have a task proceed unimpeded. Assuming normal productivities allows room for schedule compression should it become necessary.
2. Evaluate each task independently of predecessors or successors. Assume no other work existed. While professionals understand that this is not realistic, it portrays the work as unimpeded by other work. The impacts of other tasks or constraints will be calculated when the schedule tasks are linked.

3. Use consistent units to measure the duration, the accepted unit as previously noted is workdays. This prevents confusion and the requirement to convert other units to measure time.
4. Document how durations were determined, what productivity rates were used, and why the rates were modified if they were. If a productivity rate was diminished, document the logic behind the reduction. This allows the project manager to defend the calculation later if the duration comes into question. This is especially helpful in situations where a condition changed from what could have reasonable been anticipated during the bid process.

CALCULATING DURATIONS

As mentioned earlier, there are two basic methods for calculating task durations. Both methods involve productivity and a reference back to the estimate. Remember, the estimate holds the key component required for determining how long a task will take to execute. That component is its quantity. Reasonable logic suggests that the larger the quantity of the task, the longer it will take in workdays to perform. Calculating task durations with historical data from self-performed work can be relatively straightforward by one of two common methods.

Daily Production Rate Method

It is fairly common for companies to track the work they self-perform. Tracking work performance has two distinct advantages: (1) it validates the unit price carried in the estimate, and (2) it allows the scheduler to determine from actual performance records how long it will take to perform a task.

Published cost data such as that produced by the industry-leading RSMeans Company provides productivity rates for a wide variety of tasks and crews that can be used as a source or as a check and balance to the historical data.

The most common and simplest method of calculating task duration is called the *daily production rate method*. It is based on the productivity of a specific crew and the assumption that the productivity of this crew will remain constant over the life of the task provided that the conditions under which the work is performed do not change. This method recognizes that as the crew size changes the daily output will change. The daily output method is based on the simple formula:

$$D = Q \div DO$$

where
　　D　= the total duration of the task measured in workdays
　　Q　= the total quantity of the task being performed by the crew in units
　　DO　= the daily output of the task performed by the crew day in units per day.

It should be noted that the units of measure for both Q and DO must be the same. Consider the following example:

Assume a crew of two electricians has a daily output of 36 fixtures per day for installing a specific light fixture defined as Type A. The quantity takeoff portion of the estimate lists 324 as the total number of Type A fixtures.

$$\text{Duration} = 324 \text{ fixtures} \div 36 \text{ fixtures/day} = 9 \text{ days}$$

The daily production rate method is somewhat less flexible than the second method especially when the crew size is varied significantly. Productivity has a direct relationship to crew size.

Labor-Hour Productivity Method

The second method the *labor-hour productivity method* focuses less on the crew size, than the time it takes to install a single unit of the task. Most estimators use cost per unit of measure ($/SF, $/LF, $/SY, etc.) to estimate a task. There is a direct correlation between the cost in dollars and the labor-hours to perform the task. Hence, the cost can be converted to the amount of time in labor-hours regardless of the crew size. Many regard this method to be more flexible and even more accurate for scheduling. It can even be used to determine the size of the crew needed to perform the task in the allotted time.

The labor-hour productivity method uses the following formula:

$$D = TLH \div LH/d$$

where

D = the total duration of the task measured in workdays
TLH = the total quantity of labor-hours of the task being performed derived from the estimate
LH/d = the total labor-hours worked per day on the task

Consider the following example:

Historical data reveals that a single Type A fixture takes 0.5 labor-hours for an electrician to install. The total of Type A fixtures from the estimate is 324; 324×0.444 labor-hours = 143.8 ≈ 144 labor-hours.

$$\text{Duration} = 144 \text{ labor-hours} \div 16 \text{ labor-hours/day} = 9 \text{ days}$$

While both reveal that same duration in days, the labor-hour method allows the flexibility to adjust the crew size as needed.

A note about experience and where it fits in is appropriate. In roughly the same manner as the estimator adjusts costs based on actual conditions, the prebid scheduler must consider prior experience to judge how close the actual task is to ideal. For example, if a task is determined by the scheduler to be more difficult to perform in real life than it would under ideal conditions, the scheduler may want to extend the duration of the task to account for the reduced productivity.

MILESTONES AND CONSTRAINTS

One of the crucial steps in preparing the schedule is to establish intermediate goals along the critical path of the schedule. These intermediate goals are called *milestones*. Milestones act as waymarkers to measure the progress of the project as incremental portions of the work get accomplished. During actual production they allow the project manager to compare actual progress to the anticipated progress on the baseline schedule. Milestones play a big part in the monitoring process within the feedback cycle. The unique feature of milestones is that they marks the time in which a specific task or group of tasks are to be complete. It is an barometer of both time and progress.

Milestones that are triggers for the start of the next phase, or the delivery of a contract requirement should be analyzed carefully to ensure that they can be achieved under the estimated time and conditions.

10 | Bidding Strategies

I t is a fairly common goal among contractors to achieve the highest return for every project dollar—or in simpler terms maximize the profit on every project. Often, this can be done by employing a *bidding strategy* to obtain work. A bidding strategy is roughly defined as a game plan for obtaining work to bid and ultimately perform. Bidding strategies will vary from contractor to contractor, market to market and range from highly analytical mathematical formulas to the less tangible criteria such as the "gut feel."

A true bidding strategy can be defined as the skillful application of a wide range of techniques and considerations, timed correctly, to achieve the predetermined objective of securing work that has been competitively bid. This chapter will review some of the considerations the author feels has merit.

THE BUSINESS PLAN

There are few policies more critical to a successful electrical contractor, or any contractor, than how they bid on their projects. A contractor's profitability and even longevity revolves to a large degree on their ability to understand the details of a project, its inherent risk, and to estimate costs that anticipate both. Gross or prolonged misinterpretations of the complexity and risk of a project can lead to financial losses and ultimately business failures.

Bidding strategies are the execution of business development efforts that are carefully focused and consistent. A sound bidding strategy is the result of a carefully conceived and targeted *business plan*. The business plan is the direction the company will take. It is a formal statement of its business objectives such as the type of work it will target, the market in which it will compete, the skill sets and personnel it will need to achieve the goals, and why they believe the goals are attainable. Business plans are created based on the vision of the owner or management team. They develop over time and often are long term. A professional business plan has milestones as a metric for determining if objectives are being met. It also has phases or steps geared toward meeting those goals, and if nothing else, they are flexible.

Business plans are decision-making tools based on the big-picture view. They represent every aspect of planning for the growth and financial health of the contractor. Virtually every action taken by management is in furtherance of the business plan. A detailed discussion of creating a business plan is beyond the scope of this text. For information, the topic can be sourced and researched on the Internet.

COMPETITIVE BIDDING

The competitive bidding process is the primary way in which contractors obtain work. Even negotiated work most often starts as a competitive bid to get one's foot in the door. The competitive bidding process is fundamentally a decision-making process that is driven by past experiences. It is learned and refined by measuring the success on past bids using the same guidelines or strategy. It is designed to maximize returns while at the same time minimizing costs. Contractors collect, analyze, and evaluate historical bidding information and then create the optimal bidding strategy going forward. There are two key decisions in this process: (1) to bid or not to bid and (2) the appropriate markup for profit and overhead to be applied.

To Bid or Not to Bid

The first decision management makes is whether to bid a project or not to bid a project. In keeping with the business plan the project should have key features or criteria that fits into the model of where the company is headed. Most contractors have a checklist—albeit a mental checklist—of considerations. These can include questions such as:

- Is this our type of work?
- Does this fit our business plan?
- Can we capitalize the project between payments?
- Is the work in the right price range?
- Do we have sufficient resources to perform the work?
- Do we have competent supervision to manage the work?
- Do we have experience with the owner, engineer, or general contractor?
- How well did we perform on the last project of this type?
- Is the contract language punitive or exculpatory?
- Is the risk easily identified; can the impact be quantified?
- What is the likelihood the risk will occur?
- Is the project in our market and geographic area?

These questions are but a mere sampling of the considerations the estimator or management team may review prior to deciding to bid a project. It is crucial to scrutinize each bid opportunity to determine whether the project fits with the company's experience and capabilities, financial resources, risk tolerance, and the overall business plan.

Forays into new markets or types of work may be very much a growth aspect in the business plan. Contractors are advised to research these markets and the type of work carefully, and then and only then wade in one toe at a time. While growth is essential to the health and longevity of any company, it is careful, calculated, incremental growth that allows the contractor to continue to operate and manage that growth. Companies that experience exponential growth often collapse under their own inability to manage and capitalize the growth. Most contractors can recall a competitor that became "huge" only to suddenly disappear from the marketplace altogether or reemerge under a new name.

From a business perspective it would be considered less than a successful year if all of the profit made in a contractor's key market went to pay for the losses of venturing unprepared into a new market.

Application of Markup

The markup decision is fundamental to every contractor's success because it is critical to winning the work and its profitability. A key assumption in most bidding strategies is that the bid is won or lost at the markup level, and that while materials and labor pricing will vary slightly from contractor to contractor, the job will *cost* roughly the same for all bidders. For the purpose of this discussion the term *markup* will be defined as the home office overhead and profit only. While many texts include contingencies within the definition, the author has specifically omitted contingencies for the reasons stated in Chapter 7, "Indirect Costs."

Many studies have attempted to answer how the markup decision should be made, even to the point of developing models based on the premise that contractors want to maximize their profits in the competitive bid arena. These models include the general bidding model by Carr (1982) and the average bidding model by Ioannou et al. (1993) right up to the Fayek (1998) model. The takeaway point of all these models is that there is no surefire mathematical formula or algorithm that will ensure work will be won in a competitive bidding environment. They do, however, all point to the process as being an evolutionary one. That is to say that we learn from our past successes and failures and apply that learning to our future bids. There is a feedback loop that allows the contractor to adjust the bidding strategy by comparing expected outcomes to actual outcomes. Out of this comes the optimal bidding strategy that is continually refined. This learning occurs in a competitive environment where one bidder's actions affect another's actions. When a single contractor learns and adjusts, it drives others, his or her competitors, to do the same. A key to learning and improving is to understand all of the collected data available starting with home office overhead.

Home Office Overhead

As discussed in Chapter 7, "Indirect Costs," home office overhead is the cost of doing business. It is all of the ancillary costs of running and maintaining a business. For contractors to compete and share a market, there must be similarities between their companies. Rarely, if ever, does one see a large company with a fleet of new vehicles, large office staff, and large office/shop competing with a contractor that works out of his or her basement. If so, it doesn't last for long. One of these similarities is overhead expenses.

This is a textbook illustration of how one contractor's behavior affects another's. Contractors track their performance in the bidding marketplace and measure their performance against that of their competitors, who are also measuring their performance. If pricing is continually too high, then the bidder needs to make an adjustment to be more competitive. It is a self-correcting process. The first-place contractors' turn to make an adjustment is to their home office overhead. How can we do more with less? What expenses can be eliminated to reduce overhead? Streamline or downsize, whatever the term du jour.

One of the fatal flaws of the contracting world is for a company to not know what its home office overhead is. Flying blind is certainly going to end in disaster in this scenario. Accurately collecting and analyzing overhead costs is crucial to maintaining a competitive position in the marketplace. This is especially true in lean economic times. For smaller companies it may be well within the skill set of the owner to determine the overhead costs, but for larger companies, especially those companies experiencing growth, this may require specialized education and experience beyond that of the owner. To that company the author strongly urges the owner/management team to seek the advice and wisdom of a professional. It is no different than subcontracting a scope of work that one is not comfortable or qualified to perform. If there is a difference, it is that performing the work can result in failure of the task, whereas guessing at the overhead can lead to a chronic failure of the company.

Simply put, know the cost of the company's home office overhead before trying to develop a bidding strategy.

RESOURCE ANALYSIS

Another key element in developing a bidding strategy is to understand a company's strengths and weaknesses and, above all, its limitations. This is called the *resource analysis* process. A resource, simply stated, is the company's assets that can be called upon to meet the challenge posed by the work. The first resource that most often comes to mind is a company's employees.

This process should begin with an analysis of the strengths and weaknesses of the contractor's team: tradespersons, supervisors, and management. The following categories must be considered as objectively and honestly as possible:

- Individual strengths of the company's executive-level management team
- Office and field management experience with the type of work involved
- Experience of the field personnel (electricians) in the type of work
- Employee retention and satisfaction with the management
- Employee workload—steady or intermittent
- Employee satisfaction with the type of work
- Availability of additional resources if needed

In addition to the labor resources (employees) of a company, there are other resource considerations:

- Adequate historical cost records for the appropriate type of work
- Bonding capacity—single project and aggregate limits

- Financial strength of the company to capitalize work
- Supplier and vendor resources of the company
- Geographic area that can be managed effectively
- Equipment resources of the company
- Specialized equipment availability
- Reliable and timely cost control systems
- Subcontractor resources for specialized work

MARKET ANALYSIS

Most contractors tend to concentrate on only a few particular types of work. In fact, it is not uncommon for a contractor to do only one type of work. It is the contractor's experience that often dictates the market in which the company will compete. If a company and its personnel are successful with a certain type of work or projects, that tends to be their market. Occasionally, the company should pause and examine the portion of the industry in which they participate. During this process, the following items should be carefully analyzed:

- Historical trend of the market (in which the company competes)
- Current strength of the market (market vitality)
- Expected future trend of the market
- Geographic limitations of the market
- Competition from other contractors within that market
- Popularity of the market the company serves
- Risk involved in the particular market
- Typical size of projects in this market
- Expected return on investment from the market (highly specialized work generally has higher returns)

If several of these areas are experiencing stagnation or, worse, a downturn, then it is definitely appropriate to examine an alternate market in order to supplement the company's primary market. However, many managers may feel that lean economic conditions are the times when they should consolidate and narrow their market. This is justified by the Darwin-esque theory that in lean economic times some competitors cannot weather the storm and go out of business, leaving more work for those that survive. These managers are likely to expand or broaden their market only when market economics are especially strong. Situational awareness of the market in which you participate is essential for successful decision making.

Certain research should be done prior to creating a bid strategy within a particular market. The first step is to develop a source for projects to bid, such as a regional publication or online source. This is followed closely by obtaining the bid results of jobs in the prospective geographic area or market. These results should be analyzed for everything from dollar amount to the number of bidders. This is fairly easy to do in the case of public jobs, since the bid results are normally published (or at least available) from the awarding authority responsible for the project. For invitational markets, this step may present more of a challenge, since the bid results are not always divulged by the owner. Again, the bid results may appear in regional publications. The conclusions of the research may be far different than initially

There are some key aspects to developing a sound bidding strategy:

- Establish project evaluation criteria for projects you bid. Review them frequently and modify if required.
- Allow adequate time to prepare the estimate and bid. Haste is a recipe for disaster.
- Review contract documents carefully and thoroughly for discrepancies and inconsistencies in design. These can consume enormous amounts of management time and money.
- Read contract language carefully. Beware of punitive language to the contractor and language that exculpates the architect, engineer, owner, or general contractor (GC).
- Follow the business plan until it does not work—be flexible. Do not be afraid to make adjustments when required.
- Establish a bidding checklist or procedures to avoid omissions.
- Numbers do not lie; do not be afraid to withdraw your bid if you have discovered an arithmetical error.
- Do not procrastinate—solicit material and sub prices early allowing their estimator time to produce the best number.
- Develop a stable of multiple vendors and subs for the company's type of work.
- Identify, evaluate, and quantify risk accurately.
- Evaluate the client—maintain successful relationships, abandon the unsuccessful ones.
- Evaluate design professionals—bid the work of those in which you have had successful experience.
- Don't be afraid to decline to bid work that does not fit or has defective documents.
- Understand the market in which you participate.
- Bid work with a profit. Avoid clients with unrealistic expectations.
- Review and understand the schedule before committing to bid.
- Ensure that the company will have adequate resources available to do the work when needed.
- Know how this project affects the company's liquidity if awarded the project.
- Remember that there are always exceptions to the rule—especially rules *we* create!

MAXIMIZING THE PROFIT-TO-VOLUME RATIO

Once the bidding track record for the company has been established, the next step is to reduce the historical percentage of the variance. One method is to create a chart showing, for instance, the last 10 jobs on which the company was low bidder, and the dollar variance between the lowest and second-lowest bid (or median bid). Next, rank the percentage differences from 1 to 10 (1 being the smallest and 10 being the largest left on the table). An example is shown in Table 10.1

The company's "cost" ($17,170,000) is derived from the company's low bids ($18,887,000) assuming a 10% profit ($1,717,000). The "second bid" is the next lowest bid. The "difference" is the dollar amount between the low bid and the second bid (money "left on the table"). The differences are then ranked based on the

Table 10.1 Chart of Last 10 Jobs Bid

Job No.	Cost	Low Bid	Second Bid	Difference	% Difference	% Rank	Profit (Assumed at 10%)
1	$ 918,000	$ 1,009,800	$1,095,000	$ 85,200.00	9.28	10	$ 91,800
2	$1,955,000	$ 2,150,500	$2,238,000	$ 87,500.00	4.48	3	$ 195,500
3	$ 2,141,000	$ 2,355,100	$2,493,000	$ 137,900.00	6.44	6	$ 214,100
4	$1,005,000	$ 1,105,500	$1,118,000	$ 12,500.00	1.24	1	$ 100,500
5	$2,391,000	$ 2,630,100	$2,805,000	$ 174,900.00	7.31	8	$ 239,100
6	$2,782,000	$ 3,060,200	$3,188,000	$ 127,800.00	4.59	4	$ 278,200
7	$1,093,000	$ 1,202,300	$1,282,000	$ 79,700.00	7.29	7	$ 109,300
8	$ 832,000	$ 915,200	$ 926,000	$ 10,800.00	1.30	2	$ 83,200
9	$2,372,000	$ 2,609,200	$2,745,000	$ 135,800.00	5.73	5	$ 237,200
10	$1,681,000	$ 1,849,100	$2,005,000	$ 155,900.00	9.27	9	$ 168,100
	$17,170,000	$18,887,000		$1,008,000.00			$1,717,000

percentage of job "costs" left on the table for each. Figure the median difference by averaging the two middle percentages—that is, the fifth and sixth ranked numbers.

$$\text{Median \% Difference} = \frac{5.73 + 6.44}{2} = 6.09\%$$

From Table 10.1, the median percentage left on the table is 6.09%. To maximize the potential returns on a series of competitive bids, a useful formula is needed for pricing profit. The following formula has proven effective.

$$\text{Normal Profit\%} + \frac{\text{Median\% Difference}}{2} = \text{Adjusted Profit\%}$$

$$10.00 + \frac{6.09}{2} = 13.05\%$$

Now apply this adjusted profit percentage to the same list of 10 jobs, as shown in Table 10.2.

Table 10.2 Revised Profit in Application

Job No.	Company's Cost	Revised Low Bid	Second Bid	Status	Adjusted Difference	Profit (10% + 3.05%)	Total
1	$ 918,000	$ 1,037,800	$1,095,000		$ 57,200	$ 91,800 + $28,000	$ 119,800
2	$ 1,955,000	$ 2,210,100	$2,238,000		$ 27,900	$195,500 + $59,600	$ 255,100
3	$ 2,141,000	$ 2,420,400	$2,493,000		$ 72,600	$214,100 + $65,300	$ 279,400
4	$(1,005,000)	$(1,136,100)	$ 1,118,000	L	$ —	$100,500 + $30,600	$ —
5	$ 2,391,000	$ 2,703,000	$2,805,000		$102,000	$239,100 + $72,900	$ 312,000
6	$ 2,782,000	$ 3,145,100	$3,188,000		$ 42,900	$278,200 + $84,900	$ 363,100
7	$ 1,093,000	$ 1,235,600	$1,282,000		$ 46,400	$109,300 + $33,300	$ 142,600
8	$ (832,000)	$ (940,600)	$ 926,000	L	$ —	$ 83,200 + $25,400	$ —
9	$ 2,372,000	$ 2,681,500	$2,745,000		$ 63,500	$ 237,200 + $72,300	$ 309,500
10	$ 1,681,000	$ 1,900,400	$2,005,000		$104,600	$168,100 + $51,300	$ 219,400
	$15,333,000	$ 17,333,900			$ 517,100		$2,000,900

Note that the job "costs" remain the same but that the low bids have been revised. Compare the bottom-line results of Table 10.2 to those of Table 10.2 based on the two profit margins, 10% and 13.05%, respectively.

Total volume *drops* from $18,887,000 to $17,333,900.

Net profits *rise* from $1,717,000 to $2,000,900.

Profits rise while volume drops! If the original volume is maintained or even increased, profits rise even further. Note how this occurs. By determining a reasonable increase in profit margin, the company has, in effect, raised all bids. By doing so, the company loses two jobs to the second bidder (Jobs 4 and 8 in Table 10.2).

A positive effect of this volume loss is reduced exposure to risk. Since the profit margin is higher, the remaining 8 jobs collectively produce more profit than the 10 jobs based on the original, lower profit margin. From where did this money come? The money "left on the table" has been reduced from $1,008,000 to $517,100. The whole purpose is to systematically decrease the dollar amount difference between the lowest bids and the second-lowest bids.

Note that this is a theoretical approach based on the following assumptions.

- Profit must be assumed to be the bid or estimated profit, not the actual profit when the job is over.
- Bidding must be done with the same market in which data for the analysis was gathered.
- Economic conditions should be stable from the time the data is gathered until the analysis is used in bidding. If conditions change, use of such an analysis should be reviewed.
- Each contractor must make roughly the same number of bidding mistakes. For higher numbers of jobs in the sample, this requirement becomes more probable.
- The company must bid additional jobs if total annual volume is to be maintained or increased. Likewise, if the net total profit margin is to remain constant, even fewer jobs need to be won.
- Finally, the basic cost numbers and sources must remain constant. If a new estimator is hired or a better cost source is found, this technique cannot work effectively until a new track record has been established.

The accuracy of this strategy depends on the criteria listed in this chapter. Nevertheless, it is a valid concept that can be applied, with appropriate and reasonable judgment, to many bidding situations.

11 | Project Cost Control and Analysis

Some type of an internal accounting system should be used by contractors to logically gather and track the costs of a construction project as the work progresses. While this is not estimating as we know it, there are several reasons why this makes good estimating sense. First, the cost data collected can be analyzed to determine if the project was financially successful and, second, to compare actual costs to estimated costs. With this information, a cost analysis can be performed for each activity—both during the installation process and at its conclusion. This information or feedback becomes the basis for management decisions throughout the duration of the project. This historical cost data can then be used to fine-tune the unit costs in future estimates. While cost control is not the primary subject of this text, a brief overview is vital to a successful understanding of the estimating process.

COST CONTROL

The cost categories for an electrical project most often tracked are major items of construction (e.g., lighting, power, service), which can be subdivided into component activities (e.g., wire, conduit, fixtures). These activities should coincide with the system and methods of the quantity takeoff and ultimately the estimate. The accuracy of the data and its assignment to the proper cost categories is crucial to the value of the results. For instance, cost categories for an electrical project might include breakdowns such as lighting, switchgear, branch power, motor control centers and motors, and alarm systems—fire and security. Activities within one of these categories—branch power, for example—might include such items as receptacles, boxes, conduit, and wire.

The major purposes of cost control and analysis are as follows:

- To provide management with a system to monitor costs to compare to progress
- To provide cost feedback to the estimating department
- To determine the cost of change orders

- To be used as a basis for progress payment requisitions to the owner or general contractor
- To predict and manage cash flow
- To identify areas of potential cost overruns to management for corrective action
- To identify costs and time remaining to complete

It is important to establish a cost control system that is uniform—both throughout the company and from job to job. Such a system might begin with a uniform chart of accounts, which is a listing of code numbers for work activities. The chart of accounts is used to assign time and cost against work activities for the purpose of creating cost reports. A chart of accounts should have enough scope and detail so that it can be used for any of the projects that the company may win. Naturally, an effective chart of accounts will also be flexible enough to incorporate new activities as the company takes on new or different kinds of projects. Using a cost control system, the various costs can be consistently allocated. The following information should be recorded for each cost component:

- Labor charges in dollars and labor-hours, summarized from weekly time cards and distributed by task code
- Quantities completed to date, to determine unit costs
- Equipment rental costs derived from purchase orders or from weekly charges issued by an equipment company
- Material charges determined from purchase orders
- Allocation of appropriate subcontractor charges
- Job overhead items, listed separately or by component

Each component of costs—labor, materials, and equipment—is now calculated on a unit basis by dividing the quantity installed to date into the cost to date. This procedure establishes the actual installed unit costs to date.

It is also useful to calculate the percentage complete to date of the individual tasks as well as the project as whole. This is a multistep process to derive the maximum benefit from the data. Done correctly, the data can be used to forecast trends as well as track what has actually occurred.

For individual tasks the first step would be to track costs in dollars and labor-hours as well as units completed for the current period. The current period would be the week, biweekly, or even the month for projects with longer durations. This is traditionally called the *reporting period* and is set by the management for the project, or it can be a company standard that does not change. Both dollars *and* labor-hours should be tracked since labor-hours provide a more true evaluation of productivity. Cost per unit can change with pay increases or premium wages, while labor-hours remain constant across a variety of cost changes.

The second step would be to maintain a running total or *cumulative* total of the cost in dollars and labor-hours expended for the task since the start. This is called the *task-to-date* or *job-to-date* total. A comparison between the totals of the current period and the task-to-date allows the project manager to analyze whether performance is unchanged, improving, or decaying. If the productivity has decreased in the current period, then the project manager can investigate the cause and apply a corrective action plan to mitigate the decrease. The same is true for material and equipment costs. A significant fluctuation in cost, up or down, can be the impetus for the project manager to investigate.

The percentage complete for each activity is calculated based on the actual quantity installed to date divided by the total quantity estimated for the activity. For example, report results that indicate that a task has been 50% completed physically, while the costs and labor-hours total reveals that only 45% of the estimated value of the task has been expended would indicate that the task is being performed 5% more efficient than estimated. Again, this is strictly from a cost perspective without considering schedule performance.

A big-picture analysis includes an evaluation of cost for the project as a whole. This has the benefit of determining the overall success of a project from a cost perspective. In the real world of electrical contracting, some tasks on a project go as planned, some go better, and some, unfortunately, go worse. It is, however, the summation of the cost of those individual tasks that determines the financial success of a project.

It is this same process of tracking physical progress, or units completed, and comparing that against dollars and labor-hours expended to achieve the units completed. Again, if the project job cost report states that the project has expended 75% of the estimated cost with 80% physically complete, a project manager could determine that performance is better than anticipated. He or she could also predict that if this continued through the end of the project unabated, then the project as a whole would return a better profit than originally estimated.

This process of tracking and analysis is called *project control*, and while distinct from the discipline of estimating, it is inextricably interdependent with the estimating process. In the practice of project control, not only is the cost component analyzed, but also the schedule component of control must be integrated for a true snapshot of where the project lies. Remember, it is very possible to be ahead in the cost performance while still failing chronically from a schedule perspective.

The Practice of Cost Control

Any section on cost control without some discussion of the actual practice would be only half the story. Since the integration of computers in the day-to-day work of construction, virtually nothing is calculated by hand. That is especially true for the time-consuming and cumbersome tasks of project estimating, accounting, and cost control reporting. There are some fantastic software applications that are made specifically for cost control in construction. They all have a similar mechanical process to arrive at the resulting data.

As invoices for materials and equipment arrive, the project manager reviews them for validity and correctness and assigns or codes the invoice to a specific cost category. The accounting or booking staff then *post* the cost to the correct job, category, and element (materials, labor, hours, or equipment) based on the numbers coded by the project manager. Labor costs and hours are similarly posted by the payroll department or service to specific categories.

At a predetermined day or date, the accounting department produces a report for the project manager's review and analysis. The costs are then compared to the physical progress of the project. It should be noted that the majority of software applications are fairly sophisticated and can identify trends, good or bad. The key to good project control, and ultimately deriving good historical costs for estimating, is the setup of the cost categories and ensuring that all participants, even field personnel that are responsible for time cards, understand what is to be tracked and how. Finally, be sure to draft clear directions and instructions for each phase of the process. Adequate time must be spent to ensure that all who are involved (especially the foremen and supervisors) clearly understand the program and its benefits.

The analysis of categories serves as a useful management tool, providing information on a constant, up-to-date basis. Immediate attention is attracted to any center that is projecting a loss. Management can concentrate on this item in an attempt to make it profitable or to minimize the expected loss.

Frequently, items are added to or deleted from the contract via change orders. Accurate cost records are an excellent basis for determining the cost changes that will result.

A final note on the cost control system: Occasionally, job cost collection can get out of control. It is possible but highly impractical to break cost control categories down into too many categories. Major categories, with sidebars for special tasks is usually a good, sound practice. General rules for determining those special tasks include:

- Tasks with no historical data, or tasks with insufficient data to use as a basis for estimating
- Labor-intensive tasks
- Tasks that have long durations
- Tasks in which the estimated values are suspect
- Tasks with no prior experience performing

Novices have a habit of getting lost in the minutiae. Tracking too many tasks or in too much detail can lose sight of the goal. The goal is to produce sufficient data to manage the project. Remember project cost control is a tool to help management do their job.

PRODUCTIVITY AND EFFICIENCY

When using a cost control system, such as the one described here, the unit costs should reflect standard practices. Productivity should be based on a five-day, eight-hour-per-day (during daylight hours) workweek. Exceptions can be made if a company's requirements are unique. Installation costs should be derived using normal crew sizes, under normal weather conditions, and during the normal construction season.

All unusual costs incurred or expected should be recorded separately for each category of work. For example, an overtime situation might occur on every job and in the same proportion. In this case, it would make sense to carry the unit price adjusted for the added cost of premium time. Likewise, unusual weather delays, strike activity, owner/architect delays, or contractor interference should have separate, identifiable cost contributions. These are applied as isolated costs to the activities affected by the delays. This procedure serves two purposes:

- To identify and separate the cost contribution of the delay so that future job estimates will not automatically include an allowance for these "nontypical" delays.
- To serve as a basis for an extra compensation claim and/or as justification for reasonable extension of the job.

OVERTIME IMPACT

The use of long-term overtime is counterproductive on almost any construction job. In general, short periods of overtime production called *occasional overtime* have little impact on productivity. It is the *scheduled overtime* or prolonged overtime that lasts weeks or months that is referenced here. There have been numerous studies conducted that come up with slightly different numbers, but all reach the same conclusion. The longer the period of overtime, the lower the actual production rate. Table 11.1 tabulates the effects of overtime work on efficiency.

As illustrated in Table 11.1, there can be a difference between the *actual* payroll cost per hour and the *effective* cost per hour for overtime work. This is because of the reduced production efficiency with the increase in weekly hours beyond 40. This difference between actual and effective costs results from overtime work over a prolonged period. Short-term overtime work does not result in as great a reduction in efficiency and, in such cases, effective cost may not vary significantly from the

Table 11.1   Overtime

Days per Week	Hours per Day	Production Efficiency					Payroll Cost Factors	
		1 Week	2 Weeks	3 Weeks	4 Weeks	Average 4 Weeks	@1-1/2 Times	@2 Times
	8	100%	100%	100%	100%	100%	100%	100%
5	9	100	100	95	90	96.25	105.6	111.1
	10	100	95	90	85	91.25	110.0	120.0
	11	95	90	75	65	81.25	113.6	127.3
	12	90	85	70	60	76.25	116.7	133.3
	8	100	100	95	90	96.25	108.3	116.7
6	9	100	95	90	85	92.50	113.0	125.9
	10	95	90	85	80	87.50	116.7	133.3
	11	95	85	70	65	78.75	119.7	139.4
	12	90	80	65	60	73.75	122.2	144.4
	8	100	95	85	75	88.75	114.3	128.6
7	9	95	90	80	70	83.75	118.3	136.5
	10	90	85	75	65	78.75	121.4	142.9
	11	85	80	65	60	72.50	124.0	148.1
	12	85	75	60	55	68.75	126.2	152.4

Source: Reprinted from *RSMeans Estimating Handbook,* Third Edition, Wiley.

actual payroll cost. As the total hours per week are increased on a regular basis, more time is lost because of fatigue, lowered morale, and an increased accident rate.

As an example, assume a project where electricians are working 6 days a week, 10 hours per day. From Table 11.1 (based on productivity studies), the effective productive hours are 52.5 hours (87.5% of 60 hours). This represents a theoretical production efficiency of 52.5/60 or 87.5%.

Depending on the locale and day of week, overtime hours may be paid at time-and-a-half or double time. For time-and-a-half, the overall (average) actual payroll cost (including regular and overtime hours) is determined as follows:

For time-and-a-half:

$$\frac{40 \text{ reg. hrs.} + (20 \text{ overtime hrs.} \times 1.5)}{60 \text{ hrs.}} = 1.167$$

Based on 60 hours, the payroll cost per hour will be (on average) 116.7% of the normal rate at 40 hours per week. However, because the effective production (efficiency) for 60 hours is reduced to the equivalent of 52.5 hours, the effective cost of overtime is calculated as follows:

For time-and-a-half:

$$\frac{40 \text{ reg. hrs.} + (20 \text{ overtime hrs.} 1 \times 1.5)}{52.5 \text{ hrs.}} = 1.33$$

Installed cost will be 133% of the normal rate (for labor).

Therefore, when figuring overtime, the actual cost per unit of work will be higher than the apparent overtime payroll dollar increases, because of the reduced productivity of the longer workweek. These calculations are true only for those cost factors determined by hours worked. Costs that are applied weekly or monthly, such as equipment rentals, will not be similarly affected.

One last thought on overtime; many of the burdens on labor do not increase proportionally with the overtime. For example, using a fully burdened rate of $100 per hour multiplied by 1.5 for a premium rate is not correct. The estimator is directed to research burden rates carefully based on locale and project.

2 | COMPONENTS OF ELECTRICAL SYSTEMS

12 | Raceways

Raceways are channels constructed to house and protect electrical conductors—wire. This chapter contains descriptions of 11 different types of raceway, fittings, and associated installation activities. The appropriate units for measure are provided for each of these components, along with material and labor requirements and a step-by-step takeoff procedure. Cost modification factors are given when economy of scale or difficult working conditions may affect the cost of the project. The variety of raceways discussed here are typically specified in CSI MasterFormat Division 26—Electrical, section 26 05 33, Raceway and Boxes for Electrical Systems.

CABLE TRAY

A cable tray system is a prefabricated metal raceway structure consisting of lengths of tray and associated fittings (see Figure 12.1). Together, these components form a continuous, rigid, open support surface for cables. Cable tray is usually made of aluminum or galvanized steel, but is also available in polyvinyl chloride (PVC)-coated steel, fiberglass, and stainless steel.

There are three basic types of tray: ladder, trough, and solid bottom. Trays have an overall nominal depth of 4″ through 7″ and are supplied in standard lengths of 12′ and 24′. Standard widths are 6″, 9″, 12″, 18″, 24″, 30″, and 36″.

Fittings are the components that provide changes of both direction and size that are needed to bypass obstructions and vary run direction, while providing a continuous support path. Horizontal and vertical bends are offered with a selection of radii from 12″ to 48″ to match the bending capability of the cable being supported.

- **Horizontal fittings.** Horizontal bends are tray fittings that provide a change of direction in a horizontal plane (i.e., left or right). Bends are offered in 30-, 45-, 60-, and 90-degree configurations.
- **Vertical fittings.** Vertical bends are tray fittings that provide a change in direction in a vertical plane (i.e., up or down). Bends are offered in 30-, 45-, 60-, and 90-degree configurations.

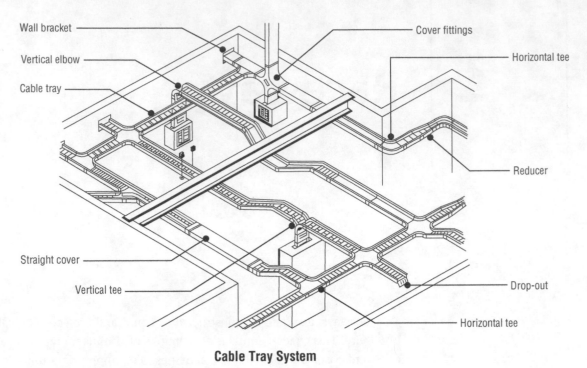

Wall bracket

Vertical elbow

Cable tray

Straight cover

Vertical tee

Cover fittings

Horizontal tee

Reducer

Drop-out

Horizontal tee

Cable Tray System

Figure 12.1 Cable Tray System

Source: *RSMeans Estimating Handbook,* Third Edition, Wiley.

- **Miscellaneous fittings.** T's, Y's, and X's are examples of other fittings available in both vertical and horizontal configurations.

Cable tray is used in many types of wire management applications. Accessories make the tray system adaptable to various configurations. The linear foot cost of cable tray excludes fittings and accessories since most configurations will vary from project to project. Accessories provide the following functions to a specific configuration: a means of exit, a means of dividing, and a means of support.

Exit Accessories
- Trough drop-out bushings
- Drop-out fittings
- Cable ties and clamps
- Conduit to tray clamps
- Cable grips

Dividing Accessories
- Barrier strips
- Straight
- Vertical inside and outside
- Horizontal

Support Accessories
- Cable tray brackets
- Trapeze hangers

- Hold-down clamps and hanger clamps
- Threaded rod
- Beam clamps

Accessories are, in most cases, not shown on the electrical plans or related details. Usually, however, a reference is made to a detail that shows a typical configuration or all of the required components are outlined in the specification section.

There are two basic types of cable tray covers—solid and vented. Solid covers are used when maximum protection of cables is required and no accumulation of heat is expected. Vented covers provide cable protection while allowing heat to dissipate from the cables. Both ventilated and solid covers protect cables from damage. In many instances, however, protection is required only in certain areas or on vertical risers subject to physical damage or for personnel protection.

Covers are available in either aluminum or steel, and finishes are offered to match the tray construction. Aluminum and pregalvanized covers 24″ wide or less come in 6′ and 12′ lengths. Covers wider than 24″ come in 6′ lengths only. Not all applications and styles of cable tray require covers, check the specifications carefully.

For straight cover, the material cost per unit includes the cover in 6′ lengths, each one including four cover clamps. The material cost for cable tray covers for horizontal and vertical bends includes four cover clamps; for tees, six cover clamps; and for crosses, eight cover clamps. Fitting covers come in sizes and widths that conform to the arrangements of the fittings.

Units of Measure. Cable tray is normally measured in linear feet (LF). The total LF must be figured for each size and depth of tray. Each unique fitting, each EA must be counted and tabulated. The same applies to each hanger style and/or support (EA). Related items, such as welded joints or field fabrication, must be identified and listed (EA).

Material Units. Various assumptions are made in all estimating standards as to what is included per unit (LF, EA, etc.) Being familiar with these assumptions and the standard being used will help to adjust material and labor prices to suit the project. In turn, estimates will be both more reliable and comprehensive.

The following items are generally included per LF of cable tray:

- Tray in standard lengths (i.e., 12′)
- One pair of connector plates every 12 LF (including bolts and nuts)
- One pair of clamp-type hangers every 12 LF
- One pair of 3/8″ threaded rods (length will be based on application).

Labor Units. The following procedures are generally included per LF of cable tray:

- Unloading/stockpiling/distribution from the delivery point
- Measuring and marking
- Installing one pair of beam clamp hangers every 12 LF
- Installing cable tray straight sections, using and including the setting up of rolling stages
- Bolting sections of tray together

- Adjusting and leveling of tray
- Laying out the work

These additional procedures are listed and estimated separately:

- Fittings and accessories
- Field fabrication
- Hanger supports
- Welded joints
- Cable

Labor adjustments are made for work over 15′ high.

Typical Job Conditions. Productivity is based on the following conditions:

- New installation
- Work plane to 15′ using rolling staging
- Unrestricted or accessible work area
- Material storage within 100′ of the installation
- Installation on the first three floors
- Minimum of 100 LF to be installed

Takeoff Procedure. Start by reviewing the electrical plans for locations and limitations of the cable tray. Refer to architectural plans and sections for dimensions—horizontal and vertical if not on the electrical plan. Ceiling heights, if applicable should also be noted. Review the typical hanger arrangement for each area being taken off: clamp type or beam type, trapeze, and so on. Mark the parameters of like types and sizes of cable tray. Separate the take off by floor, phase, wing, or some other qualifying information.

Measure or calculate the quantity of straight sections and count the various fittings for each application. The same applies to accessories. For better accuracy when performing the takeoff by hand, leave an audit trail by marking different sizes with different colors. Audit trails are routinely provided when doing the takeoff via an on screen computer system.

Transfer the quantities to the takeoff sheet, noting the height above the floor in the description. The description should contain any other pertinent information needed for accurate pricing; manufacturer, model number, finish or composition, or any defining characteristics.

Labor costs for typical installations can be sourced from Table 12.1 and modified to suit the particular application.

Cost Modifications. When tray is installed at heights over 15′, add the following percentages to labor:

15′	to	20′	high	+ 10%
20′		25′		+ 20%
25′		30′		+ 25%
30′		35′		+ 30%
35′		40′		+ 35%
Over		40′		+ 40%

Table 12.1 Installation in Labor-Hours for Cable Tray Systems

Description	Labor-Hours	Unit
Cable Tray		
Ladder Type 36" Wide	.267	LF
Elbows Vertical 36"	3.810	EA
Elbows Horizontal 36"	3.810	EA
Tee Vertical 36"	4.440	EA
Tee Horizontal 36"	5.330	EA
Drop-Out 36"	1.000	EA
Reducer 36" to 12"	2.290	EA
Wall Bracket 12"	.364	EA
Cover Straight 36"	.100	LF
Cover Elbow 36"	.320	EA

Source: Reprinted from *RSMeans Estimating Handbook*, Third Edition, Wiley.

Check the prints for penetration requirements in walls to allow the tray to go through. Adjust the estimates according to the following figures:

Up to 4 S.F. opening:

2"	Gypsum	=	0.50 LH EA
4"		=	0.57
8"		=	0.73
12"	Brick wall	=	4.00
12"	Concrete block	=	1.67
8"	Brick wall	=	2.22
5/8"	Thick drywall	=	0.33

To substitute the hanger type (if included in LF cost of cable tray) with a particular type as specified for the project, deduct the cost of the hangers used and add the type required. For example: Deduct the following from the LF material cost and labor-hours for the price of cable tray without hangers or supports.

Description	Material	Labor-Hours
6" Tray	−17%	−17%
9"	−17%	−17%
12"	−17%	−17%
18"	−16%	−17%
24"	−17%	−17%
30"	−15%	−16%
36"	−16%	−16%

Add the cost of each new type of hanger selected.

CONDUIT (TO 15' HIGH)

Conduit should be divided into the following categories: power distribution, branch lighting, branch power, and conduit for fire alarm or data cabling. Using these distinctions, the estimator can perform the takeoff based on the application category. This allows the estimator to associate the conduit with the other components (e.g., conductors, boxes, devices) in the system to arrive at the appropriate labor-hours.

The most common types of overhead conduit systems are:

- Aluminum
- Rigid galvanized steel
- Steel intermediate metallic conduit (IMC)
- Rigid steel, plastic coated
- Electrical metallic tubing (EMT)

Most plans specify the size of the conduit only. The estimator should consult the specifications for the material composition of the conduit, as this can have a dramatic impact on both material and installation costs.

Fittings are used to connect, change direction, or support conduit runs. The complexity and time required to perform particular conduit installations increase with the number of fittings required. Takeoff of conduit runs and individual fittings is performed in the same manner as with cable tray. Straight runs are tabulated by the linear foot and can be converted to quantity of pieces based on specific sales units. Fittings, for bends or couplings as well as accessories such as hangers, racks, threaded rod, and clamps are counted individually, as each (EA).

For labor purposes, the following materials are typically included per LF of aluminum, rigid galvanized, steel-intermediate, and steel-plastic coated conduit on a per-100-linear-foot basis.

- 11 Threaded steel couplings
- 11 Beam hangers
- 2 Elbows (factory sweeps)
- 4 Locknuts and 2 fiber bushings
- 2 Field thread pipe terminations
- 2 Concentric knockouts removed

For EMT:

- 11 Set screw steel couplings
- 11 Beam clamps
- 2 Field bends on 1/2″ and 3/4″
- 2 Elbows (factory sweeps) for 1″ and above
- 2 Set screw steel box connectors
- 2 Concentric knockouts removed

Labor Units. Labor costs are impacted by the number of fittings or changes in direction of the conduit installed. Productivity should be reduced for complex installations with more than the standard number of fittings per 100 LF of conduit. If the quantities of fittings exceed those in the preceding LF model, then the excess material and labor units must be added. Productivity should also be reduced for congested areas or areas that are difficult to access such as in areas of concentrated ductwork, attics, or crawl spaces.

The following procedures are generally included per LF of conduit:

- Unloading/stockpiling/distribution from the delivery point
- Setup of rolling stage
- Laying out the work
- Installing conduit, fittings, and accessories

Additional tasks and equipment should also be listed and carried in the estimate for a more comprehensive and accurate estimate. These could include:

- Staging (rental or purchase)
- Structural modifications or penetrations
- Wire and terminations
- Junction boxes
- Additional fittings or hangers
- Painting or labeling of conduit
- Fire-safing of penetrations
- Clean up

Table 12.2 provides some guidance in labor hours for the installation of various type of conduit up to 15' above grade.

Table 12.2   Installation Time in Labor-Hours for Conduit

Description	Labor-Hours	Unit
Rigid Galvanized Steel 1/2" Diameter	.089	LF
1-1/2" Diameter	.145	LF
3" Diameter	.320	LF
6" Diameter	.800	LF
Aluminum 1/2" Diameter	.080	LF
1-1/2" Diameter	.123	LF
3" Diameter	.178	LF
6" Diameter	.400	LF
IMC 1/2" Diameter	.080	LF
1-1/2" Diameter	.133	LF
3" Diameter	.267	LF
4" Diameter	.320	LF
Plastic Coated Rigid Steel 1/2" Diameter	.100	LF
1-1/2" Diameter	.178	LF
3" Diameter	.364	LF
6" Diameter	.800	LF
EMT 1/2" Diameter	.047	LF
1-1/2" Diameter	.089	LF
3" Diameter	.160	LF
4" Diameter	.200	LF
PVC Nonmetallic 1/2" Diameter	.042	LF
1-1/2" Diameter	.080	LF
3" Diameter	.145	LF
6" Diameter	.267	LF

Conduit to 15' high, includes couplings, fittings, and support.

Source: Reprinted from *RSMeans Estimating Handbook,* Third Edition, Wiley.

Takeoff Procedure. Set up the quantity takeoff sheet by size and type of conduit. Measure from the panel or source, working out to the point of termination. Do not forget to include conduit lengths for vertical changes (e.g., floor-to-floor, floor-to-above ceiling). Complete the takeoff for each diameter and type of conduit before going to another. For each category or application, summarize the quantities of each type and size conduit as well as fittings and accessories for pricing.

Cost Modifications. Add the following percentages to labor according to the height of these elevated installations:

15'	to	20'	high	+10%
20'		25'		+20%
25'		30'		+25%
30'		35'		+30%
35'		40'		+35%
40' and over				+40%

Add these percentages to the LF labor cost, but not to the material cost of the fittings. Add these percentages only to the quantities exceeding the different height levels, not to the total conduit quantities.

Where conduit runs through walls of different types of material, add appropriate cutting or coring costs. Figure the cost of each penetration by the size of conduit, composition of wall, and thickness of wall. (See Cutting and Drilling later in this chapter.)

CONDUIT FITTINGS

For estimating purposes, fittings can be defined as any component that exceeds what is considered standard in the prior section. As previously stated, the complexity of the conduit run is categorized by the quantity of fittings in the run. Simply stated, the more fittings, the more complex. In such circumstances, the estimator must tally the number of fittings and add the appropriate material, labor, and, if required, equipment cost to accurately represent the conduit run.

Locknuts are used to terminate and secure conduit endings. This procedure involves screwing the locknuts onto an existing conduit thread. Only a material price is needed. ***Note:*** Some locknuts also have a sealing ring insert for wet areas.

Bushings are used to protect wires from damage when a conduit terminates in an enclosure. There are several material variations for bushings. Among them are plastic, steel, and steel grounding bushings. Although the installation method is the same as for a locknut, several labor assumptions have been added for estimating ease. Since it is unlikely that a 10' conduit length will end at an exact location, the labor-hour cost includes field cutting and threading one conduit end.

When field threading is not practical, threadless couplings are used to join two pieces of galvanized steel conduit, and threadless connectors are used to terminate a rigid galvanized steel conduit run into an enclosure. Threaded LB's and T's are used to change the direction or split a conduit run when field bends are not practical or allowed by code.

Nipples (manufactured type) are used to join electrical enclosures or raceway where there is not enough room for standard conduit techniques. There are many variations of nipples, such as chase and offset types. Nipples are available in the same diameters as conduit. Generally, nipples are under 6″ in length.

Expansion couplings are used when the expansion and contraction of a building (or bridge) will cause damage or a loss of continuity of the conduit run. They are also used when a run exceeds a certain length—to compensate for the thermal expansion and contraction of the conduit material.

Units of Measure. Fittings are counted as each (EA) and listed by size and type of conduit material.

Material Units. There is a great variety of fittings available for each type of conduit. The most common include:

- Locknuts and bushings
- Threadless couplings and connectors
- Threaded LB's and T's
- Nipples of various lengths and diameter
- Expansion couplings
- Unions

Note: Fittings are manufactured in several types, such as explosion-proof, PVC-coated, and malleable iron. Be sure the fittings conform to the specified type of installation.

Labor Units. Different fitting types require different types of labor units:

Bushings. Generally include installing locknuts, installing bushings, and cutting and threading one conduit end.

Threadless couplings. Generally include cutting to fit two ends of rigid galvanized steel conduit (when field threading is not practical) and installing the coupling.

Threadless connectors. Generally include cutting to fit one end of a galvanized steel conduit and installing the connector.

Threaded LB's and T's. Generally include cutting and threading two ends of a conduit for LB's and three ends for T's, installing the fitting, and installing the fitting cover.

Nipples. Generally include removal of two concentric KO's and installing the nipple (including locknuts).

Expansion couplings. Generally include field threading of one conduit end, installing a bushing on one conduit end, and installing the fitting.

These basic operations can be assumed for all fittings. Explosion-proof, seal-off fittings, for example, will include the above approach plus time for packing and pouring the sealing compound.

Takeoff Procedure. In most instances, a fitting will not be shown on a print for standard conduit installations. Many of the applications of particular fittings are dictated by legal and national codes, and/or by the project's own specifications, and in some circumstances a decision by the estimator. An estimator must be aware of the electrical code, as well as the location of restrictions governing the project.

If possible, list the fittings on the same quantity sheet as the conduit. List by type and size of fitting. Count each fitting, and designate a size of the fitting. If the takeoff is done on paper plans the estimator should have some method of indicating on the plan that the fitting has been taken off.

CONDUIT IN CONCRETE SLAB

Conduit in concrete slab is used for both branch circuit and power distribution. It can also be used as a raceway for fire alarm, telephone, and data cabling. Embedded conduit is usually a very cost-effective method of raceway installation because of savings in support costs and reduced run lengths. The most common types of conduit embedded in concrete are PVC schedule 40, rigid galvanized steel, and IMC. These types of conduit are most resistant to corrosion and oxidation caused by the proximity to the concrete. Aluminum conduit is not recommended for use in concrete that contains chlorides, because it has the potential for oxidation and deterioration.

Units of Measure. Conduit in slab is taken off and listed in LF. Quantities of under slab conduits and fittings should be listed separately from above slab conduits. Fittings, such as union-type couplings, and sweeps are taken off and listed as EA. Descriptions of each should be sufficient to locate and identify the conduit run and its application.

Material Units. The following items are generally included per LF of rigid galvanized conduit, IMC conduit, or PVC conduit:

- Conduit and couplings
- Ties to slab reinforcing
- Elbows (factory sweeps)
- Locknuts and two fiber bushings
- Removal of concentric knockouts

Labor Units. The following procedures are generally included per LF of conduit:

- Unloading/stockpiling/distribution from the delivery point
- Laying out the work
- Installing conduit as previously outlined

These additional procedures are listed and extended:

- Wire
- Additional bends (over two per 100')

Takeoff Procedure. Set up the quantity takeoff sheet by size and type of conduit. Start from the panel or source and work out to the point of termination. Complete each size and type of conduit before proceeding to the next.

Cost Modifications. When multiple conduits are run parallel in slab at the same time, there is a tangible benefit to productivity. Apply the following percentage deductions to the labor-hours:

2 runs	– 5%
3	–10%
4	–15%
5	–20%
Over 5	–25%

Example #1: 6 runs of 100′ long 2″ PVC, run parallel at the same time by the same crew:

600 L.F. × .067 LH/LF	40.20 LH
Less 25%	(10.05) LH
	30.15 LH

Example #2: PVC conduit run in slab involves "stub-ups" in five locations for equipment. The total length of the runs is 100′. To adjust material and labor, it is important to remember what was included in the linear foot estimate explained in the material and labor assumptions. Next, take off the number of additional field bends or factory sweeps. Add the resulting figures to the labor-hours and material.

Total length involving stub-ups is 100 LF.

Five stubs require 10 manufactured elbows.

Total 2″ conduit runs = 100 LF.

Conduit 100 LF × .067 LH/LF	6.70 LH
Elbows 8 EA × .444 LH/EA	3.55 LH
	10.25 LH

Note: Although there are 10 elbows in the model, only 8 are included in the adjustment calculation. This is because 2 elbows were previously listed in the material per 100′ of conduit. When an elbow is added, a coupling must also be added.

CONDUIT IN TRENCH

Conduit in trench is used for power distribution and for communications. It is also commonly used for outdoor lighting applications, such as roadways, walkways, or parking lots. Conduit installation may be either direct burial or concrete encased. Rigid galvanized steel and rigid PVC conduit are usually used in trenches. For corrosive soil applications, PVC-coated rigid steel may be specified. Fiber duct or PVC conduit may also be used for the installation of a concrete duct bank.

Units of Measure. Direct burial conduit is taken off and listed by LF. Bends, fittings, spacers, and appurtenances in general are counted as separate units, EA. Excavation, sand bedding, backfill/compaction, and concrete are all calculated separately. The most common unit of measure for these tasks is cubic yard (CY). It should also be noted in the takeoff whether the excavation tasks are done by hand or by machine as this can significantly impact the cost. Precast structures such as hand holes and vaults are not included in the conduit units and must also be counted as separate units, EA. Tasks such as the excavation, bedding, backfill, and compaction as well as setting of precast structures are rarely part of the electrical contractor's work. The estimator is urged to review the documents, with focus on the specifications for clarification of the electrical contractor's scope relative to these tasks.

Material Units. The following items are generally included for rigid galvanized steel conduit:

- Conduit and threaded steel couplings
- Elbows (factory sweeps)
- Pipe terminations
- Locknuts and fiber bushings

For PVC conduit, include these items:

- Conduit and field cemented couplings
- Elbows (factory sweeps)
- Terminal adapters (male)

Labor Units. The following procedures are generally included per LF of conduit:

- Unloading/stockpiling/distribution from the delivery point
- Laying out the work
- Installing conduit
- Stabilizing conduits within the trench (if required)

These additional procedures are listed and estimated separately:

- Excavation, backfill, and compaction
- Sand bedding
- Concrete materials and placement
- Wire

Typical Job Conditions. Productivity is based on a new installation in a dry prepared trench to 4' deep. Material staging area within 200' of installation. Most often, installation of smaller-diameter conduits 2" or less can be done by as single individual, whereas large-diameter conduits may require multiple persons. Individual project circumstances can vary.

Takeoff Procedure. Set up the quantity takeoff sheet by size and type of conduit. Start from the panel or source and work out to the point of termination. Complete each size and type of conduit before proceeding to the next. The description of the task in the takeoff should indicate whether the conduit is direct burial or concrete encased.

CUTTING AND DRILLING

Cutting and core drilling of walls or floors are required in certain situations. It is fairly common in renovation work, although not exclusively. The most practical approach to this task is to identify the physical restriction requiring cutting or drilling, determine the quantity, and price it as a separate task.

Walls and floors are the most common restriction. The types that present the greatest challenge to conduit runs are usually concrete or masonry walls. Wall and floor thicknesses can range from 4″ to 24″. Since the penetration must accommodate conduit sizes from 1/2″ to 6″, the core drill must be sized larger than the conduit's outside diameter (OD). For the passage of nonround or rectangular raceways (e.g., cable tray), the opening may have to be cut versus drilled.

Units of Measure. In general, penetration through walls or floors are taken off and listed by EA. Quantities are segregated according to diameter, or in the case of rectangular openings, length by width. The composition of the wall or floor should be noted in the description since it is a major defining factor of the cost. Cutting or coring over 10′ above the finished floor should also be noted, as this can have an impact on the productivity of the task.

Material and Labor Units. The following items are generally included per hole:

- Measuring and marking
- Setup of drill equipment with available power
- Core drilling
- Cleanup and disposal of core slug

These additional items are listed and estimated separately:

- Cost of drill bits or blades
- Dust protection and personal protection equipment
- Patching and grouting
- Installation of sleeves
- Fire-safing of penetration

Typical Job Conditions. Productivity is usually based on an unrestricted area, drilling to a height of 10′ using a rotary drill or coring machine. Productivity also considers multiple penetrations per setup.

Takeoff Procedure. Set up the quantity takeoff sheet by the composition of the wall, diameter or size of the hole, and height above finish floor. Mark each penetration on the print.

Cost Modifications. When drilling at a level above 10′, add the following percentages to the labor-hours:

10′	to	15′ high—add	5%
16′	20′		10%
21′	25′		15%
26′	30′		20%
Over 30′			30%

Example #1: Twenty 2″ holes are needed in an 8″-thick concrete wall. Five conduits are at a height of 12′, 10 are at 24′, and 5 are at 35′. Adjust the labor accordingly.

5 holes	2″ @12′	=	1.8 LH × 1.05	=	9.45 LH
10	2″ @24′	=	1.8 LH × 1.15	=	20.70
5	2″ @35′	=	1.8 LH × 1.35	=	12.15
				Total	42.30 LH

Note: If only one hole is to be drilled, a two-hour minimum charge should be applied for setting up, mobilization, clean up, and demobilization.

Many specifications require the use of sleeves at cored penetrations. The sleeves have a slightly larger inside diameter than the outside diameter of the conduit and are typically steel. This concentric space between the sleeve and the conduit allow for the installation of fire-safing insulation and intumescent caulking. Tasks such as the installation of the sleeve and the fire-safing/intumescent caulking can be part of the electrical contractor's work. The estimator is urged to review the documents, with focus on the specifications for clarification of the electrical contractor's scope relative to these tasks. Fire safing and intumescent caulking at penetrations should be taken off and estimated separately.

WIRE DUCT—PLASTIC

Wire duct is designed to organize both branch and control wires within panels and enclosures. Slotted sides for wire ins/outs make this raceway accessible at 1/2″ intervals, and snap-on covers make it easy to add or change wire. Wire duct is available in standard 6′ lengths; its cross-sectional dimensions range from 0.5″ × 0.5″ to 4″ × 5″. Most wire duct is PVC and has an adhesive backing. This backing is used to hold the duct temporarily in place until it is riveted or screwed to a panel mounting plate.

Units of Measure. Wire duct is taken off and listed by the linear foot, LF for each size of wire duct.

Material and Labor Units. The following items are generally included per LF of wire duct:

- Measuring and marking
- Duct material and installation labor
- Mounting screws, drilled and tapped
- Cutting to length

These additional items/procedures are quantified, listed, and estimated separately:

- Cover
- Wire
- Marking or labeling duct
- Panel or enclosure

Takeoff Procedure. Set up the quantity takeoff sheet by the dimensions of the wire duct. Include the duct cover on the work sheet. Measure the total linear footage and round up to the next 6 LF. Mark the different sizes of duct on the plan. Complete each size and type of wire duct before proceeding to the next.

Cost Modifications. If wire duct is not field-installed in cabinets after setting, but is instead field-installed assembly-line fashion before the cabinets are mounted, then deduct the following percentages from the total labor-hours, according to the number of cabinets being worked:

1	to	3	−0%
4		6	−15%

7		9	−30%
10		15	−25%
Over 16			−30%

For example, 18 boxes are to be fabricated at benches. Thirty feet of 4″ W × 5″ H duct and cover are to be installed in each cabinet:

18 boxes × 30 LF duct/box	=	540 LF of 4″ W × 5″ H duct
	=	540 LF of cover
.160 LH × 540 LF	=	86 LH duct
.08 LH × 540 LF	=	43 LH cover
		129 LH Total
		129 × .70 = 90.3 Adjusted LH

TRENCH DUCT

Trench duct is a steel trough system set into a concrete floor, with a top cover fitted flush with the finished floor. Trench duct is used as a raceway system in concrete slabs and to feed underfloor raceways. It is also used as a flush raceway for such situations as computer rooms, language labs, shop areas, and x-ray rooms. A trench duct system is made up of individual components that allow flexibility for specific needs. The basic components are as follows:

- U-shaped trough
- Side rail assembly (standard 10′ lengths)
- Bottom plates (standard 5′ long by cover plate width)
- Cover plate of 1/4″ steel

Standard widths are 9″, 12″, 18″, 24″, 30″, and 36″, with standard depths of 2-3/8″ to 3-3/8″. All fittings come in standard duct widths. Changes in direction are made by an assortment of fittings, such as horizontal elbows, vertical elbows, crosses, and tees. The start and end accessories are end closures, risers, and cabinet connectors. All fittings must be added to material and labor calculations.

Units of Measure. Trench duct is taken off, quantified, and priced by the linear foot, LF. Fittings, supports, boxes, etc. are taken off, quantified, and priced by the unit as each, EA.

Material and Labor Units. Rather than pricing each individual component, most estimating standards have put the required components together to come up with a per LF for a complete assembly. For example, the following items are generally included per single compartment, 36″ wide trench duct:

- Two side rails
- One 36″ wide bottom plate
- One 36″ wide cover plate
- Two tack welds to a cellular steel floor
- Adjusting and leveling

These additional items are listed and estimated separately:

- Wire
- Fittings
- Access holes from duct to cellular grid
- Cutting and patching

Takeoff Procedure. Set up the quantity takeoff sheet by width and number of compartments of duct. Measure from the source of power to the end, rounding the measurement up to the next 5′. Mark the print to indicate duct and the fittings, identifying by type of fitting, width, and number of compartments. Take off accessories by type and width.

Cost Modifications. If labor-hours are based on the installation of up to 100 LF of trench duct. Adjust for quantities over 100′. This can be done by decreasing the labor-hours by the following percentages. For lengths of:

100′	to	200′	−10%
200′		300′	−20%
Over 300′			−30%

Apply these labor factors only if the lengths are located in the same general area on the same floor, and are being installed by the same crew. This applies to the labor for LF of trench duct and associated fittings.

For example, 350 LF of 24″ wide trench duct is to be installed on a first-floor area. There are five horizontal elbows within this system. Adjust the labor-hours for these quantities:

350 LF × .727 LH/LF	=	254 LH
5 EA × 5 LH/EA	=	25
		279 LH

Adjustment:

$$279 \text{ LH}(1.00 - .30) = 279 \text{ LH } (.70) = 195 \text{ LH}$$

Figure 12.2 provides a table for the labor-hours for the certain types of trench duct.

UNDERFLOOR DUCT

The purpose of underfloor duct is to make power and communication wiring available at numerous locations within a room. Used almost exclusively in office areas, the duct system is set before the concrete floor is placed and finished. Wire or cable is accessed by hand holes spaced 24″ O.C. There are two standard types of ducts; both come blank or with hand holes and are manufactured in 10′ lengths. The sizes are standard duct and, for larger wire capacities, super duct. Underfloor duct systems are available in two basic designs, single-level and two-level.

Description	Labor-Hours	Unit
Underfloor Header Duct 3-1/8″ wide	.100	LF
Underfloor Header Duct wide 7-1/4″ wide	.133	LF
Header Duct, single compartment 9″ wide	.400	LF
Header Duct, single compartment 24″ wide	.727	LF
Double compartment, 24″ wide	.800	LF
Triple compartment, 36″ wide	1.330	LF
Location market plug	.250	EA

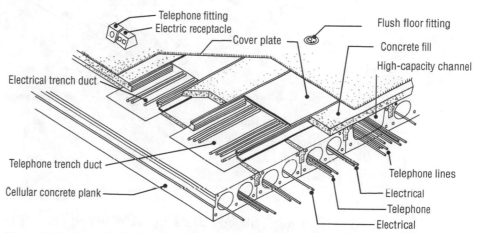

Figure 12.2   Installation Time in Labor-Hours for Cellular Concrete Floor Raceway System

Source: *RSMeans Estimating Handbook,* Third Edition, Wiley.

The *single-level duct system* involves ducts that are all mounted at the same elevation. Special junction boxes are required that accommodate one, two, or three ducts on each side. If power and communication cables run through a common junction box, a divider system must be installed to separate the two types of cables.

In the *two-level duct system*, feeder ducts are run below branch distribution ducts. Power and communication wiring can be separated more easily with this system, and junction boxes are less crowded. Figure 12.3 illustrates a single-level underfloor duct system.

Units of Measure. Underfloor duct is taken off, quantified, and priced by the LF. Final quantities should be rounded up to the next 10′ length. All fittings, junction boxes, supports, and appurtenances in general are taken off, quantified, and priced by the unit EA.

Material Units. The following items are generally included per 10 LF of underfloor duct:

- One section duct
- One coupling

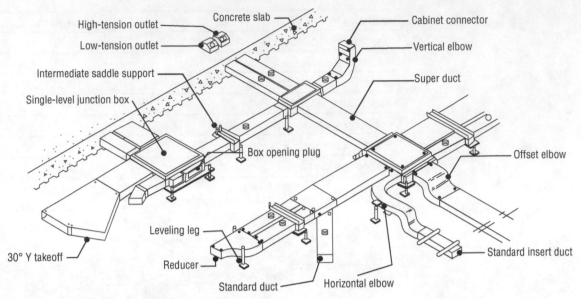

High-tension outlet — Concrete slab — Cabinet connector

Low-tension outlet — Vertical elbow

Intermediate saddle support — Super duct

Single-level junction box

Box opening plug — Offset elbow

30° Y takeoff

Leveling leg

Reducer — Standard insert duct

Standard duct — Horizontal elbow

Figure 12.3   Underfloor Raceway System

Source: *RSMeans Estimating Handbook,* Third Edition, Wiley.

These additional items are taken off and estimated separately:

- Fittings and junction boxes
- Supports and fasteners
- Adapters
- Outlets
- Wire and cable

Labor Units. The following labor procedures are generally included per 10 LF of underfloor duct:

- Unloading/stockpiling/distribution from the delivery point
- Measuring and marking
- Laying out the work
- Setting the raceway and fittings in slab or on grade
- Leveling the raceway

These additional items are taken off and estimated separately:

- Floor saw cutting
- Excavation or backfill
- Concrete placement and finishing
- Grouting or patching
- Wire
- Conduit

Takeoff Procedure. Because underfloor duct systems can vary from simple to complex, and manufacturer to manufacturer, most standards require the estimator to list all components, and to price material and labor individually. Because of the wide range and cost of fittings, junctions, and accessories, it is not safe to assume that there will be a certain number of fittings or junction boxes per 100 LF of underfloor duct.

To take off underfloor duct, set up the quantity sheet by type of duct (super or standard). Note whether or not the duct is insert type, and set it up by the on-center distance between inserts. Record the fittings separately from straight lengths of underfloor duct. Identify junction boxes as single-level or two-level. The description should indentify the different types, sizes, and partition arrangements of each junction box. An alternative would be to identify them by the manufacturer's catalog number. Most underfloor duct systems are preengineered and contain a schedule of part numbers and descriptions.

A separate part of the quantity sheet should list any accessories. This will be used to list such items as 90-degree bends, cabinet connectors, supports, offset elbows, and conduit adapters.

Take off all straight lengths of underfloor duct and mark each completed type on the plan. Transfer the lengths to the quantity sheet, and round to the next 10′ length. List and mark all junction boxes and associated fittings required, add this information to the quantity sheet. Total the quantities on all sheets.

Cost Modifications. Labor-hours are based on the installation of up to 100 LF of underfloor duct. Decrease the labor-hours by the following percentages to adjust for higher quantities.

100′	to	200′	−5%
200′		500′	−10%
Over 500′			−15%

For example, 1500 LF of 1-3/8″ × 7-1/4″ blank super duct is to be installed in the same area of an office building; adjust the labor-hours for this installation:

$$\frac{\text{Total length}}{1500} \times \frac{\text{LH per LF}}{.133} \times \frac{\text{Modifier}}{.85} = 169.6 \text{ LH}$$

Use these percentages for straight lengths only; do not deduct from the labor-hours for junctions, fittings, or accessories.

Given the variety of manufacturers, it is prudent to get material costs from a vendor or manufacturer whenever possible for the most accurate and contemporaneous pricing.

SURFACE METAL RACEWAYS

Surface metal raceway (wiremold) is installed on walls or floors. It is frequently used in applications where the surface is existing and where it would be too costly or difficult to install raceway within the wall. Surface metal raceway is available in many sizes and configurations. The smaller sizes are one-piece units and fasten to a surface with snap-on clips or straps. Wire must then be pulled in. The larger raceway consists of a base with a separate cover. The base is attached to the surface, and the cover snaps over the wires after they are put into place. Some larger sizes can be used with a divider, allowing power and telephone or other signal systems to share the same raceway.

Certain types of raceway are pre-wired with receptacles at regular spaces; in others, an assembled harness of prewired receptacles is installed in the raceway to match precut holes in the cover.

A special type of raceway is available for use on top of finished flooring. It is shallow, flat on top, and slanted on the sides to minimize tripping. Most raceway is available in 5′ or 10′ lengths, while most covers come in 5′ lengths. One-piece raceways are sold with couplings included.

Many different types of elbows, bushings, couplings, clips, adapters, boxes, and other fittings are manufactured. A catalog, readily available at most electrical supply stores, is helpful for both takeoff and installation.

Units of Measure. Wiremold is taken off, quantified, and priced by LF. It is rounded up to the nearest 10′ unit. Fittings, clips, boxes, adapters, etc. are counted as individually, quantified and priced as EA.

Material Units. The following items are generally included per 10′ of raceway:

- Sections of straight raceway and covers
- Couplings and supporting clips

All other fittings and boxes are taken off and estimated separately.

Labor Units. The following procedures are generally included in the labor-hours:

- Unloading/stockpiling/distribution from the delivery point
- Laying out the work
- Installing raceway and fastening every 10′
- Measuring and marking

These additional items are taken off and estimated separately:

- Installing wire, fittings, and boxes
- Installing device covers
- Installing raceway on walls, surfaces, or ceilings over 10′ high

Takeoff Procedure. Set up the quantity takeoff sheet according to style and size, fittings, and boxes. Take off the straight lengths of raceway for each size, then proceed to take off the fittings for each size and run. Be careful to note the correct elbow designation—horizontal, vertical, or offset. Boxes are taken off last. All of the preceding information is summarized to arrive at totals.   .

WIREWAYS

Wireways are sheet-metal troughs with hinged or removable covers. Wireways are used to house and protect electrical wires and cable. Conductors are placed after the wireway system is completed. There are two basic types of wireway based on the classification form the *National Electrical Manufacturers Association (NEMA)*. The first type is *NEMA 1*, which provides protection for wiring installations where oil, water, or dust is not an ambient concern. The second type has more stringent characteristics. It is sometimes referred to as "oiltight" and is called *NEMA 12*. It has flanged, gasketed ends and hinged, gasketed covers and is used in areas that are exposed to oil, dust, coolants, and water.

Generally, wireway is supported every 5′. However, supports every 10′ are permitted for wireway that is designed specifically for that purpose. (Vertical runs may be supported at 15′ intervals.) Wireways can be used only for exposed situations.

NEMA 1 lay-in systems come in standard sizes of 2-1/2″, 4″, 6″, 8″, 10″, and 12″ square. Standard lengths are 1′ through 5′, and 10′. Ten-foot lengths are the most common and can be assumed for estimating purposes unless otherwise specified. NEMA 12 wireways come in standard sizes of 2-1/2″, 4″, 6″, and 8″ square. Standard lengths are 1′ through 5′, and 10′.

For both NEMA 1 and NEMA 12 wireway, fittings come in the same standard widths as the tray itself. Fittings allow changes of direction for straight lengths of wireway and are grouped into four classes:

- Couplings and elbows
- Crosses and "T" boxes
- Reducers

Wireway accessories are used to terminate or support wireway installations and include items such as:

- End caps
- Panel connectors
- U connectors
- Special hangers

Units of Measure. Wireway is taken off, quantified, and priced by LF. Standard lengths are 10′, but lengths of 1′ to 5′ are available to suit the installation. Fittings, brackets, reducers, and panel connectors must be taken off, quantified, and priced individually by EA.

Material Units. The following items are generally included per 10 LF of wireway:

- One section straight raceway
- One coupling or flange kit
- Two wall bracket hangers
- One section cover

These additional items are taken off and estimated separately:

- Wire
- Fittings
- Accessories

Labor Units. The following procedures are generally included per 10 LF of wireway:

- Unloading/stockpiling/distribution from the delivery point
- Laying out the work

- Installing raceway on wall-mounted hangers up to 10′ high and spaced every 5 LF
- Installing cover

Takeoff Procedure. Set up the quantity takeoff sheet according to style and size of tray. List the fittings separately below straight sections of raceway and identify by type, size, and configuration. Take off quantities from the source and work to the termination point. Because of short sections (manufactured sizes other than 10′ long) and different sizes and types, it may be best to measure all of one size first before proceeding to another. Take off quantities, mark the print, and list all fittings; be especially careful to distinguish between vertical and horizontal fittings. Review the print once more and list the accessories (hangers, etc.). Many areas require a variety of support or hanger arrangements.

Cost Modifications. If the specifications call for fabricated hanger and support systems, the estimator should make an adjustment. To do so, deduct the following percentages from the linear foot cost for material and labor to arrive at a cost for wireway without hangers. Then add the cost of the hanger system specified.

Standard Wireway		Oiltight	
Material	−8%	Material	−4%
Labor	−4%	Labor	−4%

For example, adjust material and labor for 1,000 LF of 6″ × 6″ standard-type wireway so that hangers are not included.

	LF	×	Material $	×	Adjustment Factor	= Price without Hangers
Material:	1,000	×	$17.15	×	.92	= $15,778.00
			Unit LH			Labor-Hours
Labor:	1,000	×	.267	×	.96	= 256.32

Labor-hours are based on installing wireway to a height of 10′. To adjust labor-hours for installations over 10′, add these percentages:

11′	to	15′ high	+10%
16′		20′	+15%
21′		25′	+20%
Over 25′			+25%

Add these percentages to labor only for those lengths that are installed at the specified heights; do not add to the material price.

FLEXIBLE METALLIC CONDUIT

Flexible metallic conduit, sometimes referred to as "Greenfield" or "Flex," is a single strip of aluminum or galvanized steel, spiral wound and interlocked to provide a circular cross section of high strength and flexibility. Its flexibility, together with a continuous length, make it more cost effective than rigid conduit for use in some applications. Flexible metallic conduit can be purchased in lengths ranging from 25′ to 250′, depending on its diameter. It requires no elbow fittings or field bends for changes in direction and is easy to handle.

Flexible metallic conduit does have restrictions. It cannot be used in wet locations, hoistways, acid storage areas, hazardous locations, underground, or embedded in concrete. Sizes of flexible metallic conduit range from 3/8″ to 4″ in diameter. Flexible metallic conduit must be secured at intervals not exceeding 4.5′, as well as within 12″ of every outlet or fitting.

Conductors are pulled into flexible conduit with a pull string or "snake" in much the same manner as with rigid conduit.

Flexible conduits with a liquid tight exterior are similar to flexible metallic conduit, except that it is covered with a plastic covering. It is most often recognized by its trade name Seal-Tite® is generically called liquid tight. It is used for conduit connections to motors and raceways in machines where protection from liquids is required.

Because of the flexibility of Greenfield and Seal-Tite®, only a limited selection of fittings is necessary. The most common fittings are box connectors with either non-insulated or insulated throats. Both are available in straight and 90-degree versions. Rigid to flexible conduit couplings are also very common. Forty-five degree connectors and swivel adapters are also available for Seal-Tite®.

Units of Measure. Flexible metallic conduit is taken off by LF, and the connectors are counted as individual units, EA, according to type and size (diameter) and application (wet or dry).

Material Units. The following items are generally included in the material cost per linear foot of Greenfield and Seal-Tite®:

• The flexible metallic conduit
• The cost of a one-hole strap every 4.5′

These additional items must be taken off and estimated separately:

• Flex fittings and connectors
• Wire
• Pull string (if specified or required)

Labor Units. The following procedures are generally included per LF of flexible metallic conduit:

• Material handling on site
• Installing conduit
• Securing conduit every 4.5′ LF on mounting surface

Material and labor are based on an average of 50′ runs. These additional items are taken off and estimated separately:

• Installing wire
• Installing fittings

Takeoff Procedure. Review the electrical plans and specifications for the locations of flexible conduits and the specific application. Quantity sheets should be set up by type (wet or dry applications) and size (diameter) of flexible metallic conduit.

Also the description should include the application or any unique conditions that may impact cost. List the fittings on the same quantity sheets. There are many installation methods for flexible conduit. For example, flex used to feed branch circuits involves a takeoff procedure similar to that for rigid conduit.

Some estimators prefer to takeoff wire quantities at the same time as flex conduits. Since each size of flexible conduit has specific limitations as to the size (gauge) and number of wires that can be pulled through it, the quantity of flexible conduits can be used to calculate the wire quantity. This tabulation is done by multiplying the quantity of conduits by the number of conductors in each. Add sufficient additional length for each run for connections at either end. Summarize the totals of both flexible conduits and wire (if calculated simultaneously) and list on separate quantity sheets.

Cost Modifications. If the installation is done above suspended ceilings, the conduits may be tied in place rather than fastened with straps. This method results in a savings of both material and labor costs. Deduct 7% from the material cost and 35% from the labor in this type application.

13 | Conductors and Grounding

A conductor is a wire or metal bar with a low resistance to the flow of electric current. Grounding is accomplished by a conductor connected between electrical equipment or a circuit and the earth. In this chapter, there are 11 sections on various types of wiring and terminations, and a section on grounding. Each section begins with a definition of the component, followed by units of measure, material and labor requirements, and a takeoff procedure.

WIRE

Wire is used to conduct current from an electrical source to an electrical use. Wire is made of either copper or aluminum conductors with an insulating jacket. Copper is normally used for sizes smaller than #6, while copper or aluminum is used on #6 and larger. Wire comes with various voltage ratings and insulation materials. Some types of insulation are thicker than others and require a larger-size conduit. Aluminum conductors of equal *ampacity*, or ampere capacity, are generally larger in diameter than copper and may require larger diameter conduit.

The cost impact of using different types of wire goes beyond the cost of the wire itself. Labor productivities, and ultimately the cost for pulling various sizes and types of wire must also be considered. Another factor is that because of the composition of the wire (copper versus aluminum), the same size (gauge) wire will have different capacity ratings. Consider the following: #6 copper wire is rated at 65A, and #6 aluminum is rated at 50A. To match the 65A rating of copper wire #4 aluminum would be used, which is rated at 65A. Clearly a larger wire may impact the size conduit and also the labor to pull the larger wire.

Another consideration goes beyond the wire itself: using #6 THHN-THWN and XHHW, four wires will require 3/4″ conduit, whereas using #6 THW, four wires, requires 1″ conduit. Using #4 THHN-THWN, four wires, requires 1″, whereas using #4 THW, four wires, requires 1-1/4″. Thus, the cost of the wire, raceway, and the installation labor must be considered in arriving at the comprehensive cost when making value engineering comparisons.

Units of Measure. Wire quantity is typically expressed in units of 100 linear feet (CLF). Care must be taken to multiply the run length by the total number of individual wires to get the total length of wire. The ground wire is not always shown on the drawings, but it will be required by specification or local codes having jurisdiction.

Note: Cable and wire are not always treated the same way. Cable is an assembly of two or more insulated wires laid up together, usually with an overall jacket. While some cables are estimated in CLF units, heavier types may be quantified and estimated by linear feet (LF). The estimator should define the units by industry standard or local practice.

Material Units. Generally, the material costs include only the wire and cable. Ancillary items such as dunnage and deposits for reels should be figured separately.

Labor Units. The price for the installation of CLF of wire generally includes the following procedures:

- Setting up wire coils or spools on racks
- Attaching wire to pulling line
- Measuring and cutting wire
- Pulling wire into raceway
- Identifying and tagging

Additional items that may be required are taken off and estimated separately. Some of the more common items are:

- Terminations to breakers, panelboards, or equipment
- Splices
- Reel storage, cartage, and handling
- Work over 10′ off the finished floor

Takeoff Procedure. Wire should be taken off and separated by the following categories:

- Feeders and service entrance
- Branch lighting
- Branch power

Feeder wire is usually taken off from the feeder schedule sheet on the electrical drawings, using conduit lengths and adding footage for terminations at each end. As an example, switchboard-to-distribution panel would add 12 LF to each wire. Feeders are considered to be wire size #6 or larger. Branch circuits are #8, #10, or #12 wire size.

Wire quantities are taken off in one of two ways: by measuring each cable run, or by extending the conduit and raceway quantities times the number of conductors in the raceway. Adequate waste should be added for terminations. Keep in mind the different units of measure of wire, CLF, and raceways, LF. The estimator may be required to convert footages of raceway to conductor with the following formula:

$$(LF\ of\ Raceway \times number\ of\ conductors) \div 100 = CLF\ of\ wire$$

The appropriate amount of waste would be added to arrive at the correct CLF of wire required.

Cost Modifications. If more than three wires are being pulled at one time, there is a resulting labor efficiency that results in a labor cost savings. The following percentages can be deducted from labor for the specific groupings referenced:

4–5 wires	–25%
6–10	–30%
10–15	–35%
Over 15	–40%

If wire pull is less than 100′ in length and is interrupted several times by boxes, lighting outlets, and the like, it may be necessary to add the following lengths to each wire being pulled:

Junction box to junction box	2 LF
Lighting panel to junction box	6 LF
Distribution panel to sub panel	8 LF
Switchboard to distribution panel	12 LF
Switchboard to motor control center	20 LF
Switchboard to cable tray	40 LF

Calculating Drops and Risers. When taking off wire quantities, it is important to include the wire for drops to electrical equipment. If the heights of various electrical equipment items are not clearly stated, the following guide can be used:

	Riser (Bottom Feed)	Drop (Top Feed)
Safety switch to 100A	5′	6′
Safety switch 400–600A	4′	6′
100A panel 12 to 30 circuit	4′	6′
42 circuit panel	3′	6′
Switch box	3′	3′–6″
Switchgear	0′	8′
Motor control centers	0′	8′
1–35 kVa transformers	0′	2′
45 kVa transformers	0′	4′

Table 13.1 is a guide to labor for the installation of various type of wire in raceways. The labor-hours listed may need to be adjusted for specific applications and can vary significantly from project to project. Table 13.1 should be used as a guideline only.

Cable terminations are also a critical part of estimating wire. Table 13.2 provides guidance for the cost of various types of terminations.

Table 13.1   Installation Time in Labor-Hours for Electrical Conductors: Wire and Cable

Description	Labor-Hours	Unit
600V Copper #14 AWG	.610	CLF
#12 AWG	.720	CLF
#10 AWG	.800	CLF
#8 AWG	1.000	CLF
#6 AWG	1.230	CLF
#4 AWG	1.510	CLF
#3 AWG	1.600	CLF
#2 AWG	1.780	CLF
#1 AWG	2.000	CLF
#1/0	2.420	CLF
#2/0	2.760	CLF
#3/0	3.200	CLF
#4/0	3.640	CLF
250 kcmil	4.000	CLF
500 kcmil	5.000	CLF
1000 kcmil	9.000	CLF

Source: Reprinted from *RSMeans Estimating Handbook,* Third Edition, Wiley.

Table 13.2   Installation Time in Labor-Hours for Electrical Conductors: Cable Terminations

Description	Labor-Hours	Unit
CABLE TERMINATIONS		
Wire connectors, screw type, #22 to #14	.031	EA
#18 to #12	.033	
#18 to #10	.033	
Crimp 1 hole lugs, copper or aluminum, 600 volt		
#14	.133	
#12	.160	
#10	.178	
#8	.222	
#6	.267	
#4	.296	
#2	.333	
#1	.400	
1/0	.457	
2/0	.533	
3/0	.667	
4/0	.727	
250 kcmil	.889	
300 kcmil	1.000	
350 kcmil	1.143	
400 kcmil	1.231	
500 kcmil	1.333	
600 kcmil	1.379	
700 kcmil	1.455	
750 kcmil	1.538	

Source: Reprinted from *RSMeans Estimating Handbook,* Third Edition, Wiley.

ARMORED CABLE

Armored cable is a fabricated assembly of approved cable with an aluminum or galvanized metal enclosure. Type AC (also called BX) available in wire sizes from #12 aluminum or #14 copper through AWG (American Wire Gauge) Type AC is rated for 600V or less. Type MC (metal clad) allowed for applications at 5,000V or less. It may have a plastic overall jacket.

Both types of armored cable are used in applications where physical protection is needed and conduit is not practical or required. Armored cable is not installed in conduit, underground, or in wet locations.

One of the advantages of armored cable is protection against vermin.

Units of Measure. Armored cable is taken off and quantified in LF of cable.

Material Units. Only the cable itself is included in the takeoff.

For large-size armored cable, box connectors must be added. When using BX, the cable is connected to boxes with built-in clamps designed to hold BX. No connectors are required in this case, and labor is included in the box installation.

Labor Units. The following procedures are generally included in the installation of BX cable:

- Field preparation and cutting
- Pulling cable
- Identification and tagging of the wire

These additional procedures must be quantified and estimated separately:

- Drilling of studs or other structural members
- Wire terminations to equipment and breakers
- Work over 10' off the finished floor

Takeoff Procedure. Calculate the length of each cable run based on size and type. Quantities should be taken off and listed by the LF including adequate quantities for drops and risers. Allow for sufficient waste at terminations. Summarize quantities to arrive at totals for each type and size. Refer to Shielded Power Cables later in this chapter for special notes.

Table 13.3 provides guidance for the labor-hours to install various sizes of armored (BX) cable.

CABLE TERMINATIONS (TO 600 VOLT)

Any point at which a wire or cable starts or ends can be described as a *termination*. The estimator must determine the number of individual cable runs and provide terminations for each conductor at each end. When equipment must be set prior to pulling wire, cable terminations are not included. Switchgear, motor control centers (MCC), control panels, and the like do not include terminations in their installation cost. Terminations can be labor intensive and costly.

Table 13.3   Installation Time in Labor-Hours for BX Cable

Description	Labor-Hours	Unit
Armored cable		
600 bolt (BX), #14, 2 conductor, solid	3.333	CLF
3 conductor, solid	3.636	
4 conductor, solid	4	
#12, 2 conductor, solid	3.478	
3 conductor, solid	4	
4 conductor, solid	4.444	
#12, 19 conductor, stranded	7.273	
#10, 2 conductor, solid	4	
3 conductor, solid	5	
4 conductor, solid	5.714	
#8, 3 conductor, solid	6.154	
4 conductor, stranded	7.273	
#6, 2 conductor, stranded	6.154	

Source: Reprinted from *RSMeans Estimating Handbook,* Third Edition, Wiley.

When working with lighting and wiring devices, such as switches and receptacles, the wires and cables are in place before the fixture or device is installed. Therefore, the installation price of a receptacle, for example, includes stripping, marking, and installing wire to the terminals on the receptacle. The installation of breakers, panels, and load centers all include terminations for branch wiring.

Units of Measure. Cable terminations are taken off by counting for each conductor of a cable at each end. The unit of measure is EA.

Material Units. The following items are generally included per unit of termination:

- Termination lug
- Split bolt connector and tape (splice)
- Two-way connector and tape (splice)
- Tape (if required)

Labor Units. The following procedures are generally included per unit of termination:

- Stripping of insulation from cable
- Cleaning and application of deoxidizing compound on aluminum cable
- Installation of terminal lug or connector on wire
- Installation of terminal lug to equipment
- Taping as required

Takeoff Procedure. Start by determining and listing each cable size and type of lug on a quantity sheet. Include terminations for each conductor for each cable end. Tabulate and summarize to get totals.

SHIELDED POWER CABLES

Shielded power cables are required in systems where the voltage is 5 kV or above. In the range of 2 kV to 5 kV, the use of shielded cables is a design option that depends on the intended use or application (such as in damp conduits). Below 2 kV, shielded power cable is not required.

Shielding is the practice of wrapping semiconducting layers—either closely fitting or bonded to the inner insulation layers with no voids. The purpose is to equalize the stress within the insulation and to confine the dielectric field within the cable. Shielding extends the life of the cable, limits electromagnetic interference, minimizes surface discharges, and reduces the hazard of shock.

Voltage ratings may be from 5 kV to 69 kV, and conductor sizes range from #6 to 1,250 kcmil with copper, or to 2,000 kcmil with aluminum.

Units of Measure. Shielded cable is taken off and quantified in units of CLF as with most cable.

Material Units. The following items are generally included per CLF of cable:

- Cable
- Wire ties

Additional items such as terminations of each conductor of each shielded cable must be take off and estimated separately.

Labor Units. The following procedures are generally included for the installation of shielded cable:

- Receiving
- Material handling
- Setting up pulling equipment
- Measuring and cutting cable
- Pulling cable into conduit
- Preparation of conduit for wire

The following additional procedures must be taken off and quantified and estimated separately:

- Conduit or cable tray
- Wire terminations and splices
- Work over 10′ off the finished floor

Takeoff Procedure. Start by listing the sizes, voltage rating, and types of cables on each quantity sheet. Calculate the length of each run and list in the appropriate column. Total each column and add the appropriate footage for terminations or waste.

When taking off the cable for procurement purposes, pay particular attention to the shipping lengths ordered. Most engineers will not permit splices. For example, a standard reel length is 1,000′. If a three-conductor cable run of 400′ and is needed (i.e., 1,200′ total), it cannot be made from one reel. Additional will need to be purchased. This must be considered in the estimated quantities.

Note: Whenever possible, order cable based on actual field-measured lengths of raceways. This is a common, preferred practice to avoid conflict when scaling dimensions from plans.

CABLE TERMINATIONS (HIGH VOLTAGE)

A cable termination fastens the conductor of a wire to the devices or equipment it serves. Many low-voltage (600V and less) devices have wire attachment terminals with built-in lugs. A screw, nut, bolt, or compression device may be used to firmly and directly attach the wire. Other equipment, such as a motor, must be terminated by attaching a termination lug to the wire, usually by crimping it to the wire. The lug is then bolted or screwed to the motor's terminal. Tape or other insulation may be required.

For cables rated at 5 kV and above, the termination process is more complex. A termination lug is compressed onto the conductor, but special attention must be paid to the insulation. If the conductor's shield insulation was abruptly stopped, an electrostatic field would concentrate at the spot and, in time, cause the cable to fail. By tapering the insulation for several inches and applying layers of tape and other insulating materials, a "stress cone" is formed, which gradually equalizes the electrostatic potential over a long surface area.

In addition to stress cones, a series of skirts may be used for outdoor applications. The skirts are intended to prevent leakage from the termination along the damp surface of the cable terminations by effectively increasing the surface distance or length. Splices in high-voltage cables involve the same concerns and precautions as stress cones.

Both splices and terminations require a great deal of time and careful attention. A number of manufacturers supply kits for performing high voltage terminations.

Units of Measure. Each end of each conductor is counted as one termination (EA).

Material Units. The following items are generally included per unit:

- The termination lug
- A prepackaged kit to form a stress cone

Labor Units. The following procedures are generally included per labor unit for each termination:

- Cutting the cable to length
- Stripping the insulation
- Forming and shaping the stress cone
- Attaching the completed lug

For terminations that require testing, the cost of each test should be quantified and estimated separately.

Takeoff Procedure. The cable size, as well as the type of insulation and jacket, must be carefully noted when doing the takeoff. These characteristics can have a dramatic impact on cost.

When a stress cone is started, it should be completed without interruption. This may require the estimator to include premium time in the calculations. Count each cable by size and multiply by the number of conductors to arrive at the total number of terminations.

For low-voltage (under 600V) devices that have built-in terminals, such as outlets and switches, the labor for termination is included with the device's installation unit. No added labor calculation is necessary.

When estimating (and purchasing) termination lugs, it is important to ensure that they are compatible with the conductor material. Although some lugs are made to be suitable for either copper (CU) conductors or aluminum (AL) conductors, many are not and must be purchased accordingly.

MINERAL INSULATED CABLE

Mineral insulated cable, type MI is an assembly of one or more bare copper conductors, insulated with a highly compressed refractory mineral insulation and enclosed in a seamless metallic sheath that is both liquid- and gas-tight. The mineral insulation is magnesium oxide, and the sheath is phosphorous deoxidized copper. This type of cable can be used in hot or cold, wet or dry locations, and in hazardous areas. It can be exposed, embedded in concrete, or buried.

Mineral insulated cable is rated for 600V. Sizes range from #16 to 500 kcmil single conductor and multiple conductors up to #4-3 conductor, to #6-4 conductor, and #10-7 conductor. MI cable is also available in heating versions, in stainless steel jackets, or in #16 wire rated at a maximum of 300V.

Special fittings are used to terminate, a procedure that requires particular care. When MI cable is stripped, it must be sealed immediately with an adaptor to prevent moisture from entering the insulation. Special tools are available for use with mineral insulated cable. While MI cable does not need to be installed in conduit, conduit is sometimes used as a sleeve to protect the cable where it comes out of a concrete slab. MI cable should be supported every 6' or in accordance with the electrical code.

Units of Measure. MI cable is taken off and quantified in units of 100 CLF. Count each type of end termination and quantify by the unit, EA. Summarize the total quantity of each type.

Material Units. The following items are generally included per CLF of MI cable:

- Cable (with clips and fasteners every 6')

Each termination of MI cable should be counted and quantified as each, EA. Terminations are estimated separately.

Labor Units. The following procedures are generally included per CLF of MI cable:

- Installing the MI wire
- Fastening with bolts and clips

Terminal kits are quantified and estimated separately. Work over 10' above the floor should also be noted for accurate estimating.

Takeoff Procedure. Cable should be taken off by size, number of conductors and by type of termination (hazardous or nonhazardous). Allow 12″ or more for making terminations and connections at each end of run. Remember that the unit of measure of cable is CLF. List the lengths of each type of cable on the quantity sheets. Summarize and add the appropriate footage for terminations and waste. Count and quantify the terminations required for each size and type.

NONMETALLIC SHEATHED CABLE

Nonmetallic sheathed cable is factory-constructed of two, three, or four insulated conductors enclosed in an outer sheath. This covering is a plastic or fibrous material. Nonmetallic sheathed cable comes with or without a bare ground wire and is made up of either aluminum or copper conductors. The conductor sizes range from #14 to 4/0.

Nonmetallic sheathed cable is a category that includes a variety of individual types. Each has specific uses and code restrictions. Types of nonmetallic sheathed cable include:

- **Type NM.** Nonmetallic sheathed cable is used for dry areas in residential wiring and with restrictions in certain commercial buildings.
- **Type NMC.** Similar to type NM, but with a corrosion-resistant sheathing, can be used in damp areas.
- **Type SNM.** Shielded type NM cable is used in cable trays and raceways.
- **Type SE.** Service entrance cable has a flame-retardant, moisture-resistant covering.
- **Type USE.** Underground service entrance cable has a covering that is moisture resistant, but not flame retardant.

Units of Measure. Nonmetallic cable is calculated and quantified in units of CLF.

Material Units. The following items are generally included in the quantity of nonmetallic sheathed cable:

- Cable
- Staples, clips, or fasteners at a predetermined spacing

Additional items, such as box connectors must be counted, and quantified individually and listed as EA in the takeoff.

Labor Units. The following procedures are generally included in the labor-hours for nonmetallic cable:

- Drilling wood (or bushings in metal) studs
- Measuring and cutting to length
- Running the cable
- Staples and clips at a predetermined spacing

The following additional tasks must be taken off and estimated separately:

- Knock-outs

Table 13.4 Installation Time in Labor-Hours for Romex® Cable

Description	Labor-Hours	Unit
Nonmetallic sheathed cable 600 volt		
Copper with ground wire (Romex)		
#14, 2 conductor	2.963	CLF
3 conductor	3.333	
4 conductor	3.636	
#12, 2 conductor	3.200	
3 conductor	3.636	
4 conductor	4	
#10, 2 conductor	3.636	
3 conductor	4.444	
4 conductor	5	
#8, 3 conductor	5.333	
4 conductor	5.714	
#6, 3 conductor	5.714	
#4, 3 conductor	6.667	
#2, 3 conductor	7.273	

Source: Reprinted from *RSMeans Estimating Handbook,* Third Edition, Wiley.

- Box connections
- Terminations
- Work over 10′ off the finished floor

Takeoff Procedure. When calculating the length, allow for drops, risers, and the amount used within boxes. List each type and size required separately.

Count the box connectors, grommets, and the like and summarize to arrive at a total. It is typical practice to add 10% for waste for these items.

Table 13.4 provides guidance for the labor-hours to install various sizes of nonmetallic sheathed (Romex®) cable.

FIBER-OPTIC CABLE SYSTEMS

Fiber-optic systems use optical fiber such as plastic, glass, or fused silica, a transparent material, to transmit radiant power (i.e., light) for control, communication, and signaling applications. The types of fiber-optic cables can be nonconductive, conductive, or composite. The composite cables contain fiber optics and current-carrying electrical conductors.

Estimating the cost of a fiber-optic system is more complex than most other types of conductors. The performance of the whole system will affect the cost significantly. New specialized tools and techniques available decrease the installation cost tremendously.

Units of Measure. Fiber-optic cable is quantified in units of CLF. Other industry units of measure are the meter (m) or kilometer (km). The connectors are counted and quantified as individual units, EA.

Material Units. Generally, the material costs include only the cable. All the accessories are taken off and estimated separately.

Labor Units. The following procedures are generally included for the installation of fiber-optic cables:

- Receiving and material handling
- Setting up pulling equipment
- Measuring and cutting cable
- Pulling cable
- Work over 10′ off the finished floor

Terminations of fiber-optic cables are considered an additional task and are counted, quantified and estimated separately.

Takeoff Procedure. Cable should be taken off by type, size, number of fibers, and number of terminations required. List the lengths of each type of cable on the quantity sheets. Total each size and type and add the appropriate additional footage for terminations and waste. Remember to consider drops and risers for each individual run.

SPECIAL WIRES

The estimating category of special wire includes a number of distinct wire types, such as fixture wires, low-voltage instrumentation wire, signal cables, telephone cables, tray cables, high-performance unshielded twisted pair (UTP) communication cables, and coaxial cables. Each of these cables is designed to meet unique performance criteria for very specific applications. Fixture wire, for example, has insulation rated for the high temperatures created in lighting fixtures. Coaxial cables are designed to have specific impedances that will match the input and output equipment's impedance; this arrangement results in the maximum energy transfer of signals.

Units of Measure. The category of special wires are calculated and quantified in LF and can be extended to units of CLF. Special connectors for these cables, when required, are counted as one for each cable end, EA.

Material Units. Generally, the material costs include only the wire. All the accessories are taken off and estimated separately.

Additional items such as the following are taken off, quantified, and estimated separately:

- Special terminators (such as coaxial cable connectors)
- Strain relief grips and the like
- Special tools for connectors or stripping

Labor Units. The installation costs of special wires typically include the labor to pull and support the special wires as needed.

Additional procedures such as the following must be taken off, quantified, and estimated separately:

- Coax connector installed on cable with special crimp tool
- Strain relief devices
  - Terminations
  - Work over 10′ off the finished floor

Takeoff Procedure. List the cables by type and size on the quantity sheet. Carefully calculate cable runs for one system at a time, being sure to include drop and riser lengths in the totals.

Count fittings, terminators, or special support devices and list them on the same quantity sheet. Total each type and size as well as the fittings and allow additional wire footages for waste.

GROUNDING

In most distribution systems, one conductor of the supply is grounded. This conductor is called the *neutral* wire. In addition, the National Electrical Code (NEC) requires that a grounding conductor be supplied to connect non-current-carrying, conductive parts to ground. This distinction between the *grounded conductor* and the *grounding conductor* is important.

Grounding protects persons from injury in the event of an insulation failure within equipment. It also stabilizes the voltage with respect to ground and prevents surface potentials between equipment, which could harm both people and equipment.

Lightning protection systems are a separate concern, but are closely related to grounding in intent and practice. Lightning poses two kinds of danger. The first is the lightning strike itself. This can damage structures and distribution systems by passing a very high current for a brief time—causing heat, fire, and/or equipment failure. The second danger is from induced voltages in lines running close to the lightning's path. These pulses can be very high and can damage electrical equipment or cause injury to people.

In both systems, it is essential that a good, low-resistance ground path be provided. This can be accomplished via the use of ground rods, a ground grid, or attachment to metal pipes in contact with the earth, such as water pipes.

Note: In communities where plastic water pipes are used between the street main and a building, this grounding method is not suitable—even though copper pipes may be used for the building's interior.

Connections between ground wires and pipes, conduits, or boxes are often made with cable clamps. Permanent connections between heavy ground wires and ground rods, building structural steel, other ground wires, or lightning rods are usually made with an *exothermic* weld process. Exothermic welding requires no external source of heat or energy to weld the metals together.

Units of Measure. Ground wire is measured in LF and extended to units of CLF. Ground clamps and exothermic welds are counted as individual units and quantified in the takeoff as EA.

Material Units. The following items are generally taken off and quantified in the grounding takeoff:

- Ground rods
- Bare or insulated ground wire
- Ground clamps
- Exothermic weld metal (less molds and accessories)

Note: A different type of mold is necessary for each different type of connection. The following items are considered additional and must be taken off, quantified and estimated separately;

- Sleeves or raceways to protect ground wires
- Lightning rods and devices

Labor Units. The following procedures are taken off, quantified, and estimated as part of the grounding takeoff:

- Running ground wire and supports as required
- Ground clamps or lugs
- Exothermic weld

The following additional procedures must be considered. These are typically in addition to the grounding procedure:

- Excavation and backfill*
- Sleeves or raceways used to protect grounding wires
- Wall penetrations
- Floor cutting
- Core drilling

* Rarely are excavation and backfill part of the electrical contractor's work and are more often considered part of the Related Work section of Part 1 in the technical specifications. This work is performed by others and coordinated with the electrical contractor. The estimator is advised to confirm this by reviewing the specifications.

Takeoff Procedure. Start by listing the sizes and types of ground wires on the quantity takeoff sheets. List ground rods by diameter and length. Identify, list, and quantify each unique type of ground connection. Each shape and size of exothermic weld connection will require a separate mold. List each mold required on the takeoff sheet for material only. Calculate the length of the cables by size and allow for drops and risers and any necessary waste. Count the ground rods, clamps, and exothermic welds. Total each and summarize on the quantity sheet. For estimating purposes, add 5% to 10% to the quantity of exothermic welds.

UNDERCARPET WIRING

Branch circuits are the conductors used between the circuit protection (fuse or breaker) and plug outlets for power and lighting. Flat conductor cable—type FCC (also referred to as *Undercarpet Power Systems*)—is designed to be an accessible, flexible system for branch circuits.

Flat conductor cable is composed of three or more flat copper conductors assembled edge-to-edge in an insulating jacket. Both a top and a bottom shield are required. All components must be firmly anchored typically with adhesives, or in the case of devices, by the use of fasteners. Crossings of two types of FCC, such as telephone cable, are permitted, provided metal shield is placed between them. Transition assemblies are used to connect the FCC to conventional wiring for the home run to a breaker panel.

Type FCC cable may be used in existing buildings to rework obsolete wiring systems, and in new buildings to allow customizing fit to the tenants and users of office areas.

As the FCC cable system has gained popularity, a wide spectrum of complementary products has also become available. These products include telephone systems, flat coaxial data cable systems, and fiber-optic data link systems.

Type FCC cable must be installed under carpet squares not larger than 30" by 30". It is permitted on hard, smooth, continuous floor surfaces and on wall surfaces in metal raceways. It is not permitted outdoors or in wet locations or corrosive atmospheres and is limited to specific types of buildings.

Units of Measure. The cable and shield are calculated and taken off by LF. Fittings, taps, splices, bends, and any special connectors are counted as an individual unit, EA, and estimated separately.

Material Units. The following items are generally included for installation of undercarpet power systems:

- **Cable.** Includes flat conductor cable with bottom shield, top shield, adhesive, and tape.
- **Transition fitting.** Includes one base, one cover, and one transition block.
- **Floor fitting.** Includes one frame base kit, one transition block, and two covers (duplex/blank).
- **Tap.** Includes one tap connector for each conductor, two insulating patches, and two top shield connectors.
- **Splice.** Includes one splice connector for each conductor, two insulating patches, and two top shield connectors.
- **Cable bend.** Includes two top shield connectors.
- **Cable dead end (outside of transition block).** Includes two insulating patches.
- A special connecting tool is also required.

Labor Units. Flat conductor cable for undercarpet power systems includes placing the cable only.

Additional tasks such as the following must be quantified and estimated separately:

- Top shield
- Mark and layout of floor
- Tape primer and hold-down tape
- Patching or leveling uneven floors
- Filling in holes or removing projections from concrete slabs
- Sealing porous floors
- Sweeping and/or vacuuming floors
- Removing existing carpeting, cutting carpet squares, and installing new or reinstalling carpet squares

Transition fitting includes assembling the transition block and attaching wires. These additional tasks should be taken off and estimated separately:

- Wall outlet box for the transition fitting with cover
- Home run wires and conduit

Floor fitting includes placing the receptacle device and terminating to the cable.

Tap includes terminating the conductors with connectors. These additional tasks should be taken off and estimated separately:

- Insulating patches
- Top shield

Splice includes terminating the conductors with connectors. These additional procedures should be taken off and estimated separately:

- Insulating patches
- Top shield

Cable bend includes folding the cable at 90 degrees to change direction. These additional procedures should be taken off and estimated separately:

- Top shield

Cable dead end includes cutting the cable to length and applying insulating patches, which should be taken off and estimated separately.

Takeoff Procedure. Start by listing the cable by number of conductors on the quantity sheet. Identify and quantify the components for each fitting type, tap, splice, and bend on the quantity takeoff sheet. Each component should be quantified and estimated separately.

Start at the power supply transition fittings and survey each circuit for the components needed. Tabulate the quantities of each component under a specific circuit number. Use the floor plan layout scale to get the cable footage. Combine the count for each component in each circuit and list the total quantity in the last column.

Table 13.5   Installation Time in Labor-Hours for Undercarpet Power Systems

Description	Labor-Hours	Unit
Cable 3 Conductor #12 with Bottom Shield	.008	LF
Cable 5 Conductor #12 with Bottom Shield	.010	LF
Splice 3 Conductor with Insulating Patch	.334	EA
Splice 5 Conductor with Insulating Patch	.334	EA
Tap 3 Conductor with Insulating Patch	.367	EA
Tap 5 Conductor with Insulating Patch	.367	EA
Receptacle with Floor Box Pedestal Type	.500	EA
Receptacle Direct Connect	.320	EA
Top Shield	.005	LF
Transition Block	.104	EA
Transition Box, Flush Mount with Cover	.400	EA

Source: Reprinted from *RSMeans Estimating Handbook,* Third Edition, Wiley.

Allow approximately 5% waste on items such as cable, top shield, tape, and spray adhesive. Suggested guidelines for determining quantities:

• Equal amounts of cable and top shield should be included.
• For each roll of cable, include a set of cable splices.
• For every 1′ of cable, include 2-1/2′ of hold-down tape.
• For every three rolls of hold down tape, include one can of spray adhesive.

Adjust the final figures wherever possible to accommodate standard packaging or sales unit of the product, which should be available from the distributor.

Table 13.5 provides guidance for the labor-hours to install various components of undercarpet power systems.

UNDERCARPET TELEPHONE SYSTEMS

An undercarpet telephone system is designed to connect telephone devices to a distribution closet using undercarpet cabling. This method provides flexibility in open office situations. The general rules for estimating follow very closely with FCC power systems.

Units of Measure. The cable is taken off, quantified, and estimated per LF. Fittings, taps, splices, bends, and so on are counted as individual units, EA.

Material Units. The following items are generally included:

• Cable runs include cable as with power systems. Additional items such as tape and adhesive should be taken off, quantified, and estimated separately.
• Transition fittings include one base plate, one cover, and one transition block.
• Floor fittings include one frame base kit, two covers, and modular jacks.

Labor Units. Labor costs for the installation of cable runs include the cable only.

Additional tasks, such as the following should be taken off, quantified, and estimated separately:

- Floor marking
- Floor preparation
- Top shield
- Tape
- Spray adhesive
- Cable folds
- Conduit or raceways to transition floor boxes
- Telephone cable to transition boxes
- Terminations before transition boxes
- Floor preparation as described in power section

Transition fittings include assembly of the transition block and terminating wires only. Additional tasks, such as the following, should be taken off, quantified, and estimated separately:

- Wall outlet box with cover
- Home run wires
- Conduit and fittings

Floor fittings include placing the device, covers, modular jacks, and fastening in place. Additional tasks, such as cutting holes in carpet should be taken off, quantified, and estimated separately.

Be sure to include all cable folds when pricing labor.

Takeoff Procedure. Review the plans carefully, identify each transition. Number or letter each cable run from that fitting.

Start at the transition fitting and survey each circuit for the components needed. List the cable type, terminations, cable length, and floor fitting type under the specific circuit number.

Use the floor plan layout scale to calculate the cable footage. Add extra length (next higher increment of 5′) to cable quantities.

Combine the list of components in each circuit and summarize the total quantity on the takeoff sheet. Add the necessary 5% allowance for such items as wire, tape, bottom shield, and spray adhesive to arrive at total quantities.

Adjust the final quantities to accommodate standard packaging or sales units. Check that items such as transition fittings, floor boxes, and floor fittings—which are to be used for both power and telephone—have been priced as combination fittings, thereby avoiding duplication.

Be sure to include the marking of floors and drilling of fasteners if the fittings specified are not the adhesive type.

Figure 13.1 provides guidance for the labor-hours to install various components of undercarpet telephone systems.

Description	Labor-Hours	Unit
Cable assembly 25 pair with connectors 50′	.670	EA
3 Pair with connectors 50′	.340	EA
4 Pair with connectors 50′	.350	EA
Cable (bulk)		
3 Pair	.006	LF
4 Pair	.007	LF
Bottom shield for 25 pair cable	.005	LF
3–4 Pair cable	.005	LF
Top shield for all cable	.005	LF
Transition box, flush mount	.330	LF
In floor service box	2.000	EA
Floor fitting with duplex jack and cover	.380	EA
Floor fitting miniature with duplex jack	.150	EA
Floor fitting with 25 pair kit	.380	EA
Floor fitting call director kit	.420	EA

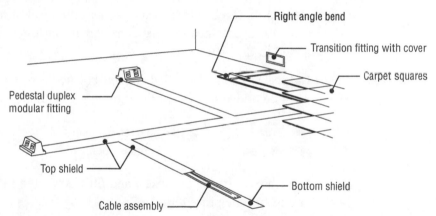

Figure 13.1   Installation Time in Labor-Hours for Undercarpet Telephone Systems

Source: *RSMeans Estimating Handbook,* Third Edition, Wiley.

UNDERCARPET DATA SYSTEMS

An undercarpet data system is designed to interconnect remote data processing terminals to the main computer by means of undercarpet wiring.

Units of Measure. The cable is taken off, quantified, and estimated per LF. Fittings, taps, splices, bends, and the like are counted as individual units, EA.

Material Units. The following items are generally included:

- Cable includes the cable only. Additional items such as tape and adhesive should be taken off, quantified, and estimated separately.

- Transition fittings include one base plate, one cover, and one transition block.
- Floor fittings include one frame base kit, two covers, and modular jacks. Additional items such as connectors should be taken off, quantified, and estimated separately.

Labor Units. Labor costs for the installation of cable runs include the cable only.

Additional tasks, such as the following should be taken off, quantified, and estimated separately:

- Conduit or raceways to transition or floor boxes
- Data cable to transition boxes
- Terminations before transition boxes
- Floor preparation as described in the power section
- Notching cable sides to make turns

Takeoff Procedure. Start the takeoff at the transition fittings and determine quantities in the same manner as the telephone system, keeping in mind that data cable does not require top or bottom shields. The data cable is simply cross-taped on the cable run to the floor fitting.

Data cable can be purchased in either bulk form or in precut lengths with connectors. If it is obtained in bulk, coaxial connector material and related labor must be added to the estimate.

Data cable cannot be folded and must be notched at 1″ intervals. A count of all turns must be added to the labor portion of the estimate. Notching requires:

90-degree turn	8 notches per side
180-degree turn	16 notches per side

Floor boxes, transition boxes, and fittings are the same as described in the power and telephone procedures.

Since undercarpet systems require special hand tools, be sure to include this cost in proportion to the number of crews anticipated to be involved in the installation.

Figure 13.2 provides guidance for the labor-hours to install various components of undercarpet telephone systems.

Description	Labor-Hours	Unit
Cable assembly with connectors 40′		
Single lead	.360	EA
Dual lead	.380	EA
Cable (bulk)		
Single lead	.010	LF
Dual lead	.010	LF
Cable notching		
90-degree	.080	EA
180-degree	.130	EA
Connectors BNC coax	.200	EA
Connectors TNC coax	.200	EA
Transition box, flush mount	.330	EA
In floor service box	2.000	EA
Floor fitting with		
slotted cover	.380	EA
Blank cover	.380	EA

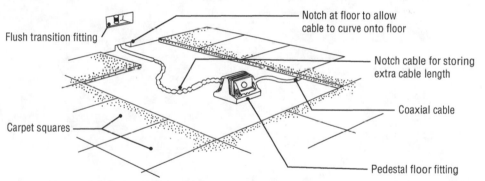

Figure 13.2   Installation Time in Labor-Hours for Undercarpet Data Systems

Source: *RSMeans Estimating Handbook,* Third Edition, Wiley.

14 | Boxes and Wiring Devices

A box is used in electrical wiring—at each junction point, outlet, and switch. Boxes provide access to electric connections and serve as a mounting for fixtures, receptacles, and switches. They may also be used as pull points for wire in long runs. A wiring device is a switch or receptacle that controls, but does not consume, electricity. This chapter has five sections that cover various types of boxes, cabinets, wiring devices, and the appropriate fasteners and hangers to secure each. Descriptions, units of measure, material and labor requirements, and a takeoff procedure are suggested for each.

PULL BOXES AND CABINETS

Pull boxes are inserted in a long run of conduit to facilitate the pulling of wire. They are also used where conduit changes direction or wires divide into different directions. Cabinets are used where wire terminates. Both come in various National Electrical Manufacturer's Association (NEMA) types to match the requirements of a specific application or environment.

Boxes and cabinets are usually made of galvanized steel, epoxy-painted steel, stainless steel, aluminum, or high-density plastic and frequently include covers. Less common materials include malleable iron, cast iron, and cast aluminum.

Units of Measure. Boxes and cabinets are taken off and quantified individually as each (EA). They are tabulated by size (dimensions), type (composition), and even by application (interior or exterior).

Material Units. The following items are generally included per unit of measure:

- Boxes
- Covers and screws

Additional items that may be required are taken off and estimated separately. Some of the more common items include:

- Fasteners to affix to mounting surface
- Support racks

- Wire terminations
- Work over 10′ off the finished floor

Labor Units. Different types of installations require different types of labor units. The following tasks are generally included for each type of application.

Wall mounting (indoor or outdoor) installations includes:

- Unloading and uncrating
- Distribution of enclosures from the stockpile
- Measuring and marking
- Drilling four anchor-type fasteners using a drill
- Mounting and leveling of boxes

Ceiling or overhead mounting includes:

- Unloading and uncrating
- Distribution of enclosures from the stockpile
- Measuring and marking
- Drilling four anchor-type fasteners using a hammer drill
- Mounting and leveling boxes from rolling staging up to 10′ above the floor

Freestanding cabinets mounting includes:

- Unloading and uncrating
- Distribution of enclosures from the stockpile
- Measuring and marking
- Drilling four anchor-type fasteners using a hammer drill
- Fastening, leveling and shimming

Telephone cabinets follow the same procedure as previously noted for each mounting condition.

Additional tasks, such as drilling or punching holes for conduit fittings, plywood backer boards, wire terminations, heights above 10′ and the assembly of multipiece cabinets should be taken off, quantified, and estimated separately.

Figure 14.1 provides guidance for estimating the labor-hours to install various types of NEMA enclosures.

Takeoff Procedure. Review the specifications for the type of the box required. List cabinets and pull boxes by NEMA type and size on the quantity sheet. Itemize and detail any special support or hanger requirements.

OUTLET BOXES

Outlet boxes made of steel or plastic are used to hold wiring devices, such as switches and receptacles. They are also used as a mount for lighting fixtures. Some outlet boxes are ready for flush mounting; others require a plaster frame. Some have plain knockouts for pipe up to 1-1/4″, while others have built-in brackets to secure the attachment of Romex® or BX cable. The capacity of outlet boxes ranges from

Description	Labor-Hours	Unit
NEMA 1		
12" Long × 12" Wide × 4" Deep	1.330	EA
20" Long × 20" Wide × 8" Deep	2.500	EA
NEMA 3R		
12" Long × 12" Wide × 6" Deep	1.600	EA
24" Long × 24" Wide × 10" Deep	3.200	EA
NEMA 4		
12" Long × 12" Wide × 6" Deep	3.480	EA
24" Long × 24" Wide × 10" Deep	16.000	EA
NEMA 7		
12" Long × 12" Wide × 6" Deep	8.000	EA
24" Long × 18" Wide × 8" Deep	20.000	EA
NEMA 9		
12" Long × 12" Wide × 6" Deep	5.000	EA
24" Long × 24" Wide × 10" Deep	20.000	EA
NEMA 12		
12" Long × 14" Wide × 6" Deep	1.510	EA
24" Long × 30" Wide × 6" Deep	2.500	EA

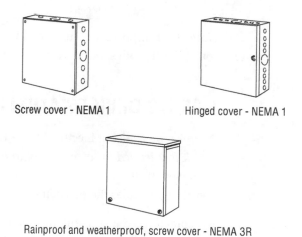

Screw cover - NEMA 1   Hinged cover - NEMA 1

Rainproof and weatherproof, screw cover - NEMA 3R

Sheet Metal Pull Boxes

Figure 14.1   Installation Time in Labor-Hours for NEMA Enclosures

Source: *RSMeans Estimating Handbook,* Third Edition, Wiley.

one to six devices. They may have brackets for direct stud mounting or "plaster ears" for mounting in existing walls.

Units of Measure. Outlet boxes are taken off and quantified individually as EA. Related accessories, such as plaster rings and covers, are also counted individually.

Material Units. The following items are generally included per unit of measure:

- Box and fastener

Labor Units. The following tasks are generally included per unit of measure for each outlet box:

- Layout and marking the box location
- Mounting the box to the stud or joist
- Material (box) handling
- Attachment of wire to the box

Additional tasks such as cutting openings for boxes in existing surfaces (old work) will require additional labor. These tasks are taken off and estimated separately.

Takeoff Procedure. Review the specifications for the type (plastic, steel, etc.) of the outlet box required. Count the individual boxes on the plans and list them on the quantity sheet. Outlet boxes can be included on the same quantity sheet as branch wiring or devices to better understand what is included in each circuit.

Cost Modifications. For large concentrations of plastic boxes in the same area, the following percentages can be deducted from labor-hours totals:

1	to	25	–0%
26		50	–15%
51		75	–20%
76		100	–25%
Over 100			–30%

Note: It is important to understand that these percentages are not applied to the total job quantities, but only in areas where concentrations reach the levels specified.

WIRING DEVICES AND LOW-VOLTAGE SWITCHING

A device by National Electrical Code (NEC) definition is *"a unit of an electrical system which is intended to carry but not to utilize electric energy."* Wiring devices include receptacles, wall switches, pilot lights, and their respective cover plates. Low-voltage switching includes the preceding plus relays, transformers, rectifiers, and controls.

Receptacles are a convenient means of connecting portable equipment and appliances to power. Receptacles are available in voltage ratings ranging from 125 to 600, and in amperages from 10 to 400. A wide variety of configurations of the contact openings prevents having a plug with a certain voltage and ampere rating inserted into the contacts of another rating. Switches, used to turn off power to lights and receptacles, are also rated for different voltages and amperages. Low-voltage switching operates at 24V through a transformer and uses relays to control 120V lighting and above. Operating at a low voltage allows the use of multiconductor #18 wire—a size much smaller than regular wire. A master control can be used to operate all lighting from one location.

Units of Measure. Wiring devices are counted as an individual unit, EA. It is quantified as the complete unit including items associated with each device, such as plates and plaster rings. An alternate method can include the box, if not included separately.

Material Units. The following items are generally included per unit of measure:

- Receptacle, wall switch, or pilot light
- Cover plates
- Outlet box and fasteners (alternate method)
- Plaster ring (if required)

For low-voltage applications, switching should include a quantity of the preceding items plus the following:

- Relays
- Transformers
- Rectifiers
- Controls

Labor Units. The following tasks are generally included per individual unit of measure:

- Stripping of wire insulation
- Attaching wire to device using terminals on the device itself
- Mounting of device in box
- Affixing cover plate

Takeoff Procedure. Quantity sheets for wiring devices should include associated accessory components as previously noted. Devices are counted by reviewing the appropriate plan. Categorize each device by its symbol or rating listed in the legend. Include all associated items necessary to complete the installation (i.e., box, cover plate, etc.).

Figure 14.2 provides guidance for estimating the labor-hours to install various types of wiring devices.

Cost Modifications. For large concentrations of devices in the same area, deduct the following percentages from labor-hours:

1	to	10	–0%
11		25	–20%
25		50	–25%
51		100	–35%
Over 100			–35%

FASTENERS

Fasteners include various types of anchors, bolts, screws, nails, rivets, and studs. Fasteners will vary based on the application, condition, and composition of the surface. Conditions vary so greatly that one cost cannot be applied to all situations. Consideration must be given to the particulars of each type installation and costs adjusted to satisfy that condition.

Units of Measure. Fasteners are taken off and quantified as individual units, EA. A complete unit includes the full assembly: bolt, nut, washer, and anchor.

Description	Labor-Hours	Unit
Receptacle 20A 250V	.290	EA
Receptacle 30A 250V	.530	EA
Receptacle 50A 250V	.720	EA
Receptacle 60A 250V	1.000	EA
Box, 4" square	.400	EA
Box, single gang	.290	EA
Box, cast single gang	.660	EA
Cover, weatherproof	.120	EA
Cover, raised device	.150	EA
Cover, brushed brass	.100	EA

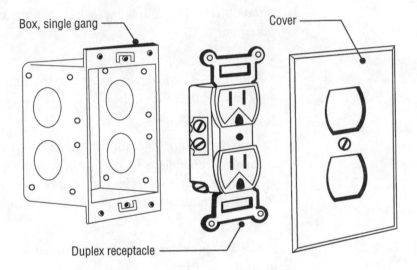

Receptacle, Including Box and Cover

Figure 14.2   Installation Time in Labor-Hours for Wiring Devices
Source: *RSMeans Estimating Handbook,* Third Edition, Wiley.

Material Units. A material unit generally is an assembly of the bolt, nut, washer, and anchor (as required) for each unit of measure.

Labor Units. The following tasks are generally included per unit of measure:

- Drilling an appropriate mounting hole for the bolt (no holes are drilled in the device being mounted)
- Inserting and setting the anchor (when appropriate)
- Threading on and tightening the fastener

Takeoff Procedure. Quantities can be determined by using the previous take off quantities for boxes, cabinets, panels, and so on. On the quantity sheet separate the different types of fasteners based on the assumptions made for mounting. This takeoff is frequently performed on the same sheet as that for the device being mounted, once all quantities of the devices have been totaled.

Cost Modifications. Many installation labor costs include anchors, with several installed at one time in one area. Labor costs must be added for unusually difficult situations, or reduced for a large quantity installed at one time in the same area. These judgments must be left to the discretion of the estimator.

HANGERS

For the purposes of the text, a *hanger* is defined as a means of suspending or supporting electrical raceway installations. As mentioned in previous chapters, hangers are often included in the linear foot cost of a raceway at predetermined intervals. Variations from these standard intervals become considerations for the estimator and may require that labor and material costs be adjusted.

Hangers can be grouped into two distinct categories—those that are shop manufactured and those that are field fabricated. Manufactured hangers are designed to support by attaching to an existing building structure without modification. Examples are one- or two-hole straps, beam clamps, or riser clamps. A field-fabricated hanger is the manual combination of two or more components to provide support. A trapeze-type hanger is a typical example. To assemble a trapeze-type hanger, the following components are required:

- Channel steel (Unistrut®)
- Threaded rod
- Nuts, bolts, and washers
- Support steel
- Beam clamps or fasteners
- Conduit straps, brackets, connectors, and so on

Each of these items requires that a material and labor cost be added to the estimate. It can be costly to disregard the cost of hangers or to assume that the linear foot cost of raceways will be accurate enough. This is particularly true on projects such as process piping, where entire networks of conduit racking systems are used.

Units of Measure. Unique hangers must first be estimated as mini-systems or assemblies to arrive at a cost for material and labor for each specific configuration. The hanger assemblies are then counted and quantified as individual units, EA.

Material Units. The material price for hangers generally includes only the hanger itself with the particular components required to complete the assembly.

Labor Units. The assembly of each component and installation of the completed unit to the building structure are generally the only tasks included in the labor-hours.

Takeoff Procedure. Review the plans and specifications carefully for details on hanger arrangements. Consult the conduit quantity takeoff for determining the quantity of each type of hanger based on the spacing specified. This count is based on the LF total for each size of raceway, divided by the specified distance between supports. If the hangers and supports are field fabricated, quantify and estimate each as subassemblies to get the total cost of each fabricated hanger. Then divide the LF total of raceway by the specified distance between supports.

If the hangers are supporting groups of conduits, then count each hanger individually. Set up the quantity sheet by the type and size of hanger. Take off the quantities, including risers and clamps if required.

15 | Starters, Boards, and Switches

Starters are electric controllers that accelerate a motor from rest to running speed; they are also used to stop the motor. Boards serve as a mounting for electric components and/or controls. Switches are devices used to open, close, or change the connection of an electric circuit. This chapter contains 18 sections and includes components ranging from panelboards and circuit breakers to motor control centers and meter sockets. Each of these components is defined, supplied with the appropriate units of measure, followed by material and labor requirements and the procedures involved in the takeoff.

CIRCUIT BREAKERS

Circuit breakers are used in general distribution and in branch circuits to protect the wires and equipment downstream from current overload. Circuit breakers are rated in amperes and are capable of interrupting their rated current at their rated voltage. There are several types of circuit breakers: magnetic trip only, molded case, current-limiting (which includes three coordinated current-limiting fuses), and electronic trip. An important parameter in selecting a breaker is its fault current interrupting capacity. Several different grades of breakers are available in most sizes, for different uses and conditions. The common *ampere interruption capacity (AIC)* ratings are 10,000 AIC; 22,000 AIC; 42,000 AIC; and 65,000 AIC. Their application is determined by the amount of energy available at the breaker's line side during a short-circuit fault.

Breakers are available in several voltages from 120 to 600, in one, two, or three poles. Enclosed breakers are available up to 2,000A. Breakers come in frame sizes of 100A, 225A, 400A, 1000A, 1200A, and 2000A. Several different ampere ratings use the same frame size.

Example: 400A frame size has ratings of 125A, 150A, 175A, 200A, 225A, 250A, 300A, 350A, and 400A. When estimating circuit breakers, it is important to use correct National Electrical Manufacturer's Association (NEMA) designations pertaining to areas of use.

Units of Measure. Circuit breakers are taken off and quantified by the individual breaker as each (EA).

Material Units. The cost for each circuit breaker generally includes fasteners.

Labor Units. The following procedures are generally included per unit:

• Unloading and distribution
• Measuring and marking circuit breaker location
• Drilling four anchor-type fasteners using a drill
• Installing the circuit breaker
• Connecting and phase marking wire on primary and secondary terminals

Additional items that may be required are taken off and estimated separately. Some of the more common items are:

• Primary or secondary conduit
• Backboards
• Wire

Takeoff Procedure. Set up the quantity sheet by circuit breaker type, voltage, number of poles, NEMA classification, and ampere rating. Count the circuit breakers, listing each one under the proper designation on the quantity sheet.

Cost Modifications. If several circuit breakers are installed in a common location by the same electrician or crew, an adjustment can be made. Deduct the following percentages from the labor units:

0–5	– 0%
5–10	– 5%
10–20	–10%
21–30	–15%
Over 30	–18%

CONTROL STATIONS

Control stations are enclosures for mounting control switches and pilot lights. (Also see Control Switches later in this chapter.)

Standard-duty control stations usually come completely assembled as an enclosure with control switches and legends. They have a capacity for up to three units and use either push buttons or selector switches. Pilot lights are also available with these stations.

Control stations are assembled in a wide variety of configurations and NEMA classifications. Standard-duty stations with NEMA enclosures have one to three positions or holes, while NEMA 4, 7, and 9 have one or two positions.

Other control stations (without switches) are heavy-duty oiltight NEMA 4 and 13 with up to 30 positions; or watertight, corrosion-resistant NEMA 4, 4X, or B for up to 25 units.

NEMA 1 enclosures come with prepunched knock-outs. However, other NEMA enclosures should be specified for hole sizes and arrangement at the time of purchase.

Oiltight and watertight enclosures may be ordered with controls pre-mounted. This saves considerable labor-hours in the field.

Units of Measure. Each station or enclosure is taken off and quantified as an individual unit, EA.

Material Units The following items are generally included per control station unit of measure:

- Enclosure
- Switch assembly (for some stations)
- Wire marker for each contact
- Legend tag for station

Additional items that may be required are taken off and estimated separately. Some of the more common items are:

- Branch wire
- Branch conduit
- Switches, pilot lights, and legends for each station

Labor Units. The following tasks are generally included per control station:

- Installing enclosure
- Mounting switches if required
- Testing

Additional items that may be required are taken off and estimated separately. Some of the more common items requiring additional labor are:

- Installing branch wire to enclosure
- Terminations of control wires
- Marking of control wires
- Installing conduit to enclosure
- Cutting or punching mounting hole

Takeoff Procedure. Review the plans and specifications for the type and location of the stations. Set up the quantity sheet by type of control station. Remember to note how many devices are in each enclosure and in each type of enclosure. A common preference is for this work to be taken off simultaneously with motors and motor controls. This allows the estimator to complete the components for each motor.

Table 15.1 provides guidance for the labor to install motor control systems.

FUSES

Fuses are used to interrupt a circuit in the event of overload. Fuses have a fusible link through which the current must flow. If too much current flows, the resulting heat melts the link. This stops the flow of current. There are many different types

Table 15.1   Installation Time in Labor-Hours for Motor Control Systems

Description	Labor-Hours	Unit
Heavy Duty Fused Disconnect 30 Amps	2.500	EA
60 Amps	3.480	EA
100 Amps	4.210	EA
200 Amps	6.150	EA
600 Amps	13.330	EA
1200 Amps	20.000	EA
Starter 3-pole 2 HP Size 00	2.290	EA
5 HP Size 0	3.480	EA
10 HP Size 1	5.000	EA
25 HP Size 2	7.270	EA
50 HP Size 3	8.890	EA
100 HP Size 4	13.330	EA
200 HP Size 5	17.780	EA
400 HP Size 6	20.000	EA
Control Station Stop/Start	1.000	EA
Stop/Start, Pilot Light	1.290	EA
Hand/Off/Automatic	1.290	EA
Stop/Start/Reverse	1.510	EA

Source: Reprinted from *RSMeans Estimating Handbook,* Third Edition, Wiley.

of fuses, such as plug, cartridge, and bolt-on. Fuses are designed with fast-acting or time-delay links. Both types will have the ability to react quickly to short-circuit currents.

Some plug fuses have an Edison screw base, which is the same as a medium base lamp. Others have a type "S" base, which requires an adapter screwed into the medium base socket. The purpose of these adapters is to discourage the use of fuses larger than the size appropriate to the existing wiring. Plug fuses are made up to 30A at 125V. Cartridge fuses are available in many types, including renewable and nonrenewable, dual element, time delaying, and current limiting. Most are rated 250V or 600V and ampacities up to 600. Bolt-on (cartridge) fuses are available to 6,000A. Cartridge fuses are categorized by class designations such as Class H, K, RK5, J RK1, or L. These designations indicate the interrupting rating (i.e., 10,000A, 100,000A, or 200,000A) and the performance characteristics of the fuse.

Units of Measure. Fuses are taken off and quantified by the individual unit of measure, each, EA. The quantity sheet should organize fuses by voltage rating, current rating, and class.

Material Units. The fuse itself is generally the only item included.

Labor Units. The following procedures are generally included per unit:

- Replacing the fuse if necessary
- Installing new fuses
- Testing with a continuity tester

The labor units previously noted are based on the receiver or fuseholder being mounted in an easily accessible location, in an unlocked enclosure.

Takeoff Procedure. Review the plans and specifications for the type and location of the fuses. Set up the quantity sheet by type of fuse. The quantity sheet should list by ampere, voltage, and type or class. Be sure to count one fuse for each line (or phase) being protected.

LOAD CENTERS

A *load center* is a specialized type of panelboard used principally for residential applications. These panels are designed for lighter sustained loads than those used in industrial and commercial applications.

Load centers are available from 100A to a maximum 200A rating in several configurations: that is, 120/240V, one phase, three wire; 120/208V, three phase, four wire; 240V, one phase, two wire; 120/240V, three phase, four wire; and 240V, three phase, three wire. Load centers generally use plug-in circuit breakers. The single-pole breakers range from 15A to 50A; double-pole breakers from 15A to 100A; and three-pole breakers from 15A to 100A. Two- and three-wire switched neutral and ground fault interrupting styles are also available.

Lighting and appliance load centers are limited to a maximum of 42 overcurrent devices in each cabinet.

Units of Measure. Load centers are taken off and quantified as individual units, each EA, by type and size. While standard breakers may be included, care must be taken to ensure that all special service breakers are noted and priced as additional items.

Material Units. The following items are generally included as a complete unit, EA, of a load center:

- Breakers
- Box
- Cover
- Trim

Labor Units. The following tasks are generally included per complete unit, EA, of a load center:

- Receiving, unloading, and uncrating
- Measuring and marking walls
- Drilling four lead anchor-type fasteners using a hammer drill
- Mounting and leveling panel
- Preparation and termination of feeder cable to lugs or main breaker
- Testing and load balancing
- Marking panel directory

Additional items that may be required are taken off and estimated separately. Some of the more common items are:

- Modifications to enclosure (extra breakers)

- Structural supports
- Additional lugs
- Plywood backboards
- Painting, lettering, or name plates

Note: Knock-outs are included in the price of terminating conduit runs and need not be added to the load center costs.

Takeoff Procedure. Review the plans and specifications for the type and location of load centers. Set up the quantity sheet by ampere, voltage, and type or class. When estimating load centers, list panels by size and type. List breakers in a separate column of the quantity sheet, and define by phase (poles) and ampere rating.

METER CENTERS AND SOCKETS

Meter sockets are enclosures designed to receive the plug-in utility watt meters that monitor a customer's power usage. For multiple tenants in commercial or residential buildings, meter centers may be used to monitor and distribute a single service entrance cable to two or more different users. This is accomplished with multiple sockets.

A *meter center* may have a main breaker, a fusible disconnect, or a remote protection device. If it feeds two or more branch services, a meter center will usually have individual meter breakers. Meter centers are available in a variety of configurations for one to six meters and can be bussed together for larger groupings. Bus capacities normally range from 125A to 800A, though meter centers with a capacity to 1200A may be special ordered.

The two styles of enclosures commonly used are surface-mounted and semiflush. NEMA 1 (general, indoor) and NEMA 3R (outdoor, rainproof) are available with provisions for top or bottom feed of the main service cable.

Meters are owned and installed by the utility company providing the service. Therefore, no cost allowance needs to be made for material or labor.

Units of Measure. Meter sockets are taken off and quantified as individual units, EA. A meter center is a unit of measure, EA, for multiple sockets.

Material Units. The following items are generally included per unit of measure:

- Meter socket or a meter center
- Wall anchor fasteners

Additional items that may be required are taken off and estimated separately. Some of the more common items are:

- Main circuit breaker or fused disconnect
- Tenant (or meter) circuit breakers
- Conduit and fittings
- Termination lugs

Labor Units. The following procedures are generally included per meter socket or meter center:

- Receiving and handling
- Leveling, alignment, and fastening to surface
- Legend identifications

Additional items that may be required are taken off and estimated separately. Some of the more common items are:

- Conduit
- Cables
- Terminations
- Breakers
- Testing

Takeoff Procedure. For both meter centers and meter sockets review the plans and specifications carefully for the specific requirements of each. Quantity sheets should differentiate between meter sockets and meter centers. Both are taken off by counting and are listed according to the following characteristics:

- For meter sockets: Count and list all sizes and types of the sockets required, based upon rating, application and type. Summarize the quantities of each.
- For meter centers: Count and list each meter center, noting its bus capacity, number of meter sockets, and type of enclosure. Take off the size of the main breaker or fusible disconnect switch. Note the size of each tenant breaker. Summarize the quantities of each.

Figure 15.1 provides guidance for labor-hours to install meter centers.

MOTOR CONTROL CENTER

Many commercial buildings maximize efficiency by grouping electrical controls into centralized locations. Motor control centers (MCCs) serve this need by providing a structure for mounting a variety of motor starters, auxiliary controls, and feeder tap units. An MCC is a collection of motor control equipment and bus bars assembled in a series of steel-clad enclosures. The enclosure, including the bus, is called a *structure* and will take any combination of starters up to 72″ high. Some structures allow for mounting starters on both the front and back sides.

A starter combination MCP, FVNR (Full Voltage NonReversing) with control transformer size 1 is 12″ high; thus, six size 1 starters could fit into one structure. Size 2 is 18″ high, so four size 2, or two size 2 with three size 1 would be acceptable. Size 3 is 24″, 4 is 30″, 5 is 48″, and 6 is 72″. Assorted combinations of these (and other) controls can be assembled. Also, many structures may be *bussed* together to expand the MCC. The enclosures are commonly available in NEMA classes:

NEMA 1:	General purpose, indoor
NEMA 2:	Drip-proof
NEMA 3R:	Rainproof
NEMA 12:	Dust-tight and drip-tight, indoor

Description	Labor-Hours	Unit
Main Breaker Section		
800 amps	17.800	EA
1200 amps	21.000	EA
1600 amps	23.500	EA
Main Section 100 amps		
6 meters	26.700	EA
8 meters	30.800	EA
10 meters	33.300	EA

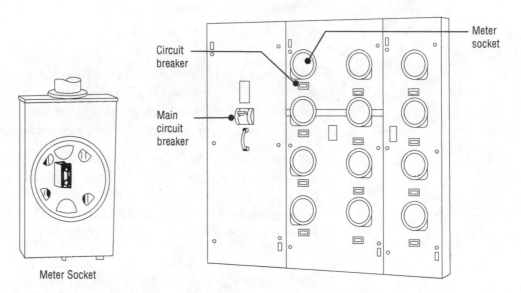

Meter Socket

Figure 15.1   Installation Time in Labor-Hours for Metering Switchboards
Source: *RSMeans Estimating Handbook,* Third Edition, Wiley.

There are two class designations for MCC wiring. In *Class I*, all internal control wiring is done in the field. In *Class II*, the control circuits are factory-wired to terminal strips. There are also three NEMA-type designations: type A, B, or C. The type indicates how the control and power cables are terminated (i.e., wired to terminal blocks or direct to the devices).

Internal bus bars are usually aluminum, but copper is also an option frequently specified. Material pricing can vary significantly with structure size, NEMA class, and type. Generally, an MCC is ordered as a complete assembly, including starters, breakers, doors, etc. The MCC can, however, be modified with field-installed components.

Units of Measure. MCCs are taken off by the section and quantified as one unit, EA. All components are identified and listed in the takeoff. Summarize the quantities of each section for pricing.

Material Units. The following items are generally included in the takeoff:

- MCC, complete as required
- Thermal heaters as required

- Fuses as required
- Wire markers

Labor Units. The following procedures are generally included in the labor-hours:

- Receiving, uncrating, and handling
- Setting equipment in place
- Leveling and shimming
- Anchoring to mounting surface
- Cable and wire terminations (power only, not control cables)
- Cable identification (power)

Additional items that may be required are taken off and estimated separately. Some of the more common items are:

- Rigging for larger structures
- Equipment pad
- Steel channels embedded or grouted in concrete
- Conduit and wire
- Testing

Takeoff Procedure. Review the plans and specifications carefully for type of structure, NEMA class, bus size, copper or aluminum bus, and NEMA type. List each MCC by the engineer's designation, and every section in each MCC. List (per MCC) all components that will be ordered in the MCC. Note all components that are factory installed so that there is no duplication of labor-hours. Summarize the quantities of each MCCs for estimating.

Figure 15.2 provides guidance for labor-hours to install motor control centers.

MOTOR CONTROL CENTER COMPONENTS

MCC components are modular devices that plug into an MCC structure's bus. They are generally specified, and purchased with the MCC as a complete assembly. Occasionally, components are purchased individually to be installed into an existing MCC. When ordering components to fit an existing installation, it is necessary to order the same make and model of components. The products of different manufacturers may not be interchangeable.

Some of the commonly used components are combination starters with either circuit breakers or fused disconnects; reversing and two-speed motor starters; feeder circuit breakers; main circuit breakers; lighting panelboards; and transformers.

The National Electrical Code (NEC) wiring class and type must also be noted. For example, with type A components, all field control wiring must be terminated directly to the device. This unit is less expensive to purchase but more costly to install. A type B component has all intrawiring brought out to terminal blocks (only the power cable is terminated directly to the starter). This approach simplifies field installation and maintenance repairs. Type C components are similar to type B except that terminal blocks are also provided for the load terminals in starters size 3 and under. Installation labor is about the same for types B and C, but maintenance and service are easier to perform.

Description	Labor-Hours	Unit
Structures 300 amps 72" high	10.000	EA
Structures 300 amps 72" high (back-to-back type)	13.300	EA
Starts Class 1 B		
Size 1	3.000	EA
Size 2	4.000	EA
Size 3	8.000	EA
Size 4	11.400	EA
Size 5	16.000	EA
Size 6	20.000	EA
Pilot light wiring in starter	.500	EA
Push-button wiring in starter	.500	EA
Auxiliary contacts in starter	.500	EA

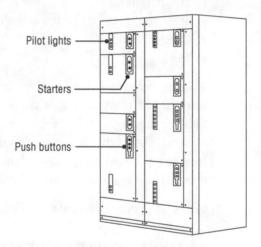

Figure 15.2 Installation Time in Labor-Hours for Motor Control Centers

Source: *RSMeans Estimating Handbook,* Third Edition, Wiley.

While some general material costs are available from reference books, by far the most reliable pricing source will be the specified manufacturer or manufacturer's representative.

Units of Measure. MCC components are taken off and quantified as each unit, EA, with careful attention to the use and to specifications as well as manufacturer of the MCC.

Material Units. The following items are generally included per MCC component unit of measure:

- Starter or other components as required
- Thermal heater as required
- Fuses as required
- Wire markers

Additional items that may be required are taken off and estimated separately. Some of the more common items are:

- MCC structure
- Any wire or raceways to the structure

Labor Units. The following tasks are generally included per MCC component:

- Receiving and handling of equipment
- Installing components in structure
- Connecting power wires to components
- Wire identification
- Installing three thermal heaters where required

Additional items that may be required are taken off and estimated separately. Some of the more common items are:

- Testing
- Control cables

Installation labor is based on the MCC being in place, ready to receive components.

Takeoff Procedure. Carefully review the plans and specifications for all components. Components are taken off and quantified with specific attention to categorize each component by location, size, type, class, and service. Summarize similar type of components for pricing.

MOTOR CONNECTIONS

A *motor connection* is the means by which the power leads are terminated to a motor. From the estimator's perspective, the motor terminations must also account for connecting the flexible conduit and fittings from the rigid conduit to the motor terminal box at its use point.

Flexible conduit to motors serves four functions. First, it can absorb the normal motor vibrations that would, over time, fatigue and crack rigid conduit connections. Second, since the vibration does not loosen flexible conduit as it would rigid parts, the ground path continuity is not disrupted. Third, most motor installations require that the motor be movable so that its shaft can be aligned with the shaft of the driven device. Finally, there is a distinct labor savings in connecting a flexible conduit to a motor box over field-bending rigid conduit to fit.

Many motors have a winding space heater. That cable is often routed and counted as a separate connection.

Units of Measure. Each cable to a motor is quantified as one unit of measure, EA for estimating purposes.

Material Units. This is actually a mini-system. The size of each component is predetermined by the standard size of cable and raceway used by each horsepower motor. This approach saves the estimator from having to dig through a conduit plan to find the size of cable and conduit servicing the motor.

The following items are generally included for motor connectors:

- Flex connectors
- Flexible metallic conduit
- Wire markers
- Termination for cable conductors

Labor Units. The following procedures are generally included in the labor-hours for motor connectors:

- Cutting flexible metal conduit
- Installing two flex connectors
- Routing wire through flex
- Terminating leads to motor
- Identifying leads

Additional items that may be required are taken off and estimated separately. Some of the more common items are:

- Mounting or placing motor (only when done by electricians)
- Disconnect switch
- Motor starter
- Motor controls
- Conduit and wire
- Checking motor rotation

Takeoff Procedure. When taking off and quantifying motor connections, it is advisable to list connections on the same quantity sheet as motors and/or motor starters. Differentiate take off quantities by horsepower rating of the motors. Count the number of motors to be connected and transfer the results to the quantity sheet. Summarize quantities from the takeoff sheet for estimating.

Cost Modifications. If several motors are to be connected in a concentrated area by the same electrical crew, the following percentages can be deducted from labor only:

1	to	10	– 0%
11		25	–20%
26		50	–30%

MOTORS

Motors are machines for converting electrical energy into mechanical energy. They are used in appliances, elevators, and all kinds of machinery. Motors are available in a multitude of sizes, voltages, types, and enclosures. Motor sizes vary from fractional HP, single phase, to as high as 2000 HP, 3 phase. Voltages typically range from 120V to 4160V AC. The most common are fractional 110V to 200 HP 600V. The most common housings are drip-proof and totally enclosed.

A range of DC motors operating at voltages from 12V DC to 250V DC are used for some special applications. Their use is, however, quite rare. Motors are not usually placed by the electrician, but rather by the trade furnishing the equipment they power, although this is not an exclusive policy. The estimator is directed to the technical specification section, in particular the *Related Work* in Part 1 for confirmation.

Units of Measure. Motors are taken off by the individual unit of measure, EA.

Material Units. Included in the material units is the motor only. Additional material items that may be required are taken off and estimated separately. Some of the more common items are:

- Pulleys
- Sheaves
- Couplings

Labor Units. The following procedures are generally included per motor:

- Receiving, unloading, and handling
- Setting the motor in place (when done by electricians)
- Bolting or fastening
- Preliminary alignment

Additional items that may be required are taken off and estimated separately. Some of the more common items are:

- Connecting the motor
- Rigging of larger motors
- Testing for rotation
- Final alignment (usually by millwrights)
- Heater connections (temporary power) for motors in storage

Takeoff Procedure. Carefully review the plans and specifications for each type motor. Consult other discipline drawings and specifications list all sizes of motors by horsepower, number of phases, and voltage. Motors should be listed at the same time as starters, disconnects, controls, and motor connections. Before pricing motors, check to see who furnishes and sets them in place.

MOTOR STARTERS AND CONTROLS

Motor starters and across-the-line starters are manufactured in a variety of sizes and types. The most common types of across-the-line starters are the following:

- Motor-starting switch with no overload protection
- Single throw switch with overload protection
- Magnetic switch with thermal overload protection

A *motor-starting switch* is simply a tumbler rotary, lever, or drum switch. It does not provide any protection against overload or inadequate voltage. It is used for nonreversing duty in small-size motors up to 2 HP, and for reversing duty up to 10 HP.

- Control switches (unless included in material price line)
- Control transformer
- Plywood backboard for mounting (if required)

Labor Units. These procedures are generally included in the labor-hours for starters:

- Unloading, uncrating, and handling of starters
- Measuring and marking
- Four lead anchor-type fasteners for mounting
- Termination of power (line) and load side cables
- Wire identification
- Installing three thermal overload heaters

Additional items such as testing may be required and is taken off and estimated separately.

Takeoff Procedure. It is not uncommon for motor starters to be furnished by someone other than the electrical contractor. Often, the starters are furnished as part of a mechanical or heating, ventilating, and air conditioning (HVAC) package. Check the specifications carefully to ensure that the starters are not by others.

Take off and list the starters by size, voltage, NEMA enclosure, and type on the quantity sheet. Count each starter and list it in the proper grouping, rather than searching for all of one type and size before proceeding to another. Frequently, starters are listed on a schedule. In addition the schedule will indicate whether integral controls are to be included. If the control is remotely located, the starter should be priced separately. If it is not, then an integral control system must be added to that starter's cost. Summarize quantities for estimating.

Note: The selection and purchasing of overload heaters for starters must be based on the actual nameplate amps of the motors installed. It is not prudent to purchase overload heaters from general charts or tables.

Cost Modifications. When several magnetic starters are installed in a common location by the same electrician or crew, the following percentages can be deducted from the labor units:

0–5	–0%
5–10	–5%
11–20	–10%
21–30	–15%
Over 30	–18%

Table 15.2 provides guidance for labor-hours to install motor starters.

Table 15.3 provides guidance for labor-hours to install motor starters in hazardous areas.

The labor-hours guidance provided is based on new construction and an installation height of 5' above grade.

Table 15.2 Installation Time in Labor-Hours for Starters

Description	Labor-Hours	Unit
Starter 3-Pole 2 HP Size 00	2.290	EA
5 HP Size 0	3.480	EA
10 HP Size 1	5.000	EA
25 HP Size 2	7.270	EA
50 HP Size 3	8.890	EA
100 HP Size 4	13.330	EA
200 HP Size 5	17.780	EA
400 HP Size 6	20.000	EA
Control Station Stop/Start	1.000	EA
Stop/Start, Pilot Light	1.290	EA
Hand/Off/Automatic	1.290	EA
Stop/Start/Reverse	1.510	EA

Source: Reprinted from *RSMeans Estimating Handbook,* Third Edition, Wiley.

Table 15.3 Installation Time in Labor-Hours for Starters in Hazardous Areas

Description	Labor-Hours	Unit
Circuit Breaker NEMA 7 600 Volts 3 Pole		
50 Amps	3.480	EA
150 Amps	8.000	EA
400 Amps	13.330	EA
Control Station Stop/Start	1.330	EA
Stop/Start Pilot Light	2.000	EA
Magnetic Starter FVNR 480 Volts 5 HP Size 0	5.000	EA
25 HP Size 2	8.890	EA
Combination 10 HP Size 1	8.000	EA
50 HP Size 3	20.000	EA
Panelboard 225 Amps M.L.O. 120/208 Volts		
24 Circuit	40.000	EA
Main Breaker	53.330	EA
Wall Switch, Single Pole 15 Amps	1.510	EA
Receptacle 15 Amps	1.510	EA

Source: Reprinted from *RSMeans Estimating Handbook,* Third Edition, Wiley.

CONTACTORS

A magnetic contactor is basically a relay with heavy-duty contacts. An electromagnetic coil is energized to close the contacts and hold them closed. Magnetic contactors are used to switch heating loads, capacitors, transformers, or electric motors. They do not have overload protection and are available from one to five poles (i.e., separate circuits). Contactors are controlled by a wall switch or push-button station. Magnetic contactors are similar to magnetic motor starters, with the exception that they do not contain overload protection devices.

Lighting contactors are a combination of a magnetic contactor with either fuse clips or a circuit breaker. They are used to control lighting and electric resistance heating loads.

Units of Measure. Contactors are taken off and quantified as an individual unit, EA.

Material Units. The following items are generally included per unit:

- Contactor
- Enclosure
- Fasteners

Additional items that may be required are taken off and estimated separately. Some of the more common items are:

- Conduit
- Wire
- Control switches
- Tags or labels

Labor Units. The following tasks are generally included per individual unit:

- Receiving and handling
- Mounting

Additional tasks that may be required are taken off and estimated separately. Some of the more common tasks requiring more labor are:

- Control and power terminations
- Control switches

Takeoff Procedure. Contactors should be taken off and listed by size, voltage, and amp ratings. The quantity sheet should also note the enclosure types required. Summarize the quantities of each type for estimating.

RELAYS

A *relay* is a control device that takes one input to a coil and operates a number of isolated circuit contacts in a control scheme. An electromagnetic coil is used to operate an armature, which holds the contacts. The relay's contacts may be set up in the following ways: open at rest (normally open), and closing when energized; or closed at rest (normally closed), and opening when energized. A relay may have from 1 to 12 or more poles.

A wide variety of relay configurations is available. Some common variations are time-delay relays (either delay on or delay off); latching relays (mechanically held needing two coils: one for on, another for off); and AC or DC relays.

Coils for relays can be specified for either AC or DC circuits and for voltages from 6 to 480. Although relays can be purchased in an enclosure such as a NEMA 1 general-purpose indoor, they are usually bought individually for assembly into a larger enclosure.

Units of Measure. Relays are taken off and quantified as individual units of units, EA.

Material Units. Generally, only the relay itself is included in the units of measure. Additional items that may be required are taken off and estimated separately. Some of the more common items are:

- Enclosure housing the relay
- Fasteners for mounting the relay

Labor Units. The following tasks are generally included per unit of measure:

- Installing the relay in an enclosure
- Connecting the wires (coil only)

Items that may require additional labor-hours are taken off and estimated separately. Some of the more common items are:

- Mounting the enclosure
- Fastening to the wall
- Conduit
- Wire
- Control terminations

Takeoff Procedure. Plans and specifications should be reviewed carefully for the location, quantity and type of relays needed. Relays should be taken off and quantified on the same takeoff sheet as the item they control. The estimator should exhibit care in pricing relays as there are many types and costs can vary considerably.

Carefully note the coil voltage, contact voltage and amp rating, number of poles, and type of operation. List each relay on the quantity takeoff and summarize the totals of each.

PANELBOARDS

Panelboards are used to group circuit switching and protective devices into one enclosure. Lighting and appliance panelboards for residential applications are referred to as load centers. (Refer to Load Centers earlier in this chapter.)

Panelboards consist of an assembly of bus bars and circuit protective devices housed in a metal box enclosure. The box provides space for wiring and includes a trim plate with a cover.

Although panelboards are available as fused models, styles using molded case circuit breakers are far more common. Panelboards are available in a variety of configurations, including AC styles up to 600V and DC styles up to 250V. The AC styles can be single- or three-phase. Models for AC up to 240V, with mains up to 600A and branch circuits up to 100A are commonly available for lighting and power distribution. Panelboards for motor starters and power distribution may have mains up to 1200A and branch breakers up to 600A.

Panelboard enclosures are also available to suit a variety of NEMA classifications, including NEMA 1, 3, and 12.

Units of Measure. Panelboards are taken off and listed by size (ampacity), type, voltage, and fault current capacity. Individual panelboards are considered a unit of measure, EA. Takeoff quantities should record the breakers with each panelboard for accurate pricing. It should be noted that some standard panelboard and breaker assemblies are available as pre-assembled units with predetermined breaker sizes and quantities.

Material Units. The following items are generally included per panelboard:

- Panelboard or panelboard assembly
- Fasteners for mounting
- Cover

Labor Units. The following tasks are generally included for each panelboard installation:

- Receiving and handling
- Measuring and marking
- Drilling four lead anchor-type fasteners using a hammer drill
- Mounting and leveling panel
- Preparation and termination of feeder cable to lugs or main breaker
- Branch circuit identification
- Lacing using tie wraps
- Testing and load balancing
- Marking panel directory
- Installing cover and trims

Items that may require additional labor-hours are taken off and estimated separately. Some of the more common items are:

- Modification to enclosure (additional breakers)
- Structural supports
- Additional lugs
- Plywood backboards
- Painting, lettering, or plaques
- Adding breakers
- Field assembly
- Terminating branch wiring

Notes: Knock-outs are included in the price of terminating conduit runs and need not be added to the panelboard costs. Material and labor prices generally include breakers.

Takeoff Procedure. Review the plans and specifications for quantity, type, size, and locations of individual panelboards. Panelboards should be taken off and quantified according to size, type, and voltage. List breakers in a separate column of the quantity sheet, and define by phase (poles) and ampere rating. Summarize quantities of each for estimating.

Table 15.4 provides guidance for labor-hours to install a variety of panelboards.

Table 15.4 Installation Time in Labor-Hours for Panelboards

Description	Labor-Hours	Unit
Panelboard 3-Wire 225 Amps Main Lugs		
38 Circuit	22.220	EA
4 Wire 225 Amps Main Lugs 42 Circuit	23.530	EA
3 Wire 400 Amps Main Circuit Breaker		
42 Circuit	32.000	EA
4 Wire 400 Amps Main Circuit Breaker		
42 Circuit	33.330	EA
3 Wire 100 Amps Main Lugs 20 Circuit	12.310	EA
4 Wire 100 Amps Main Circuit Breaker		
24 Circuit	17.020	EA

Source: Reprinted from *RSMeans Estimating Handbook,* Third Edition, Wiley.

PANELBOARD CIRCUIT BREAKERS

A panelboard *circuit breaker* is a molded-case plastic breaker (with one, two, or three poles). Two styles are available: plug-on and bolted (to the bus bars). There are models for either DC service up to 250V or for AC service up to 600V. Main circuit breakers come in many sizes ranging from 100A up to 1200A and branch circuit breakers from 15A to 800A.

A major consideration in selecting and pricing circuit breakers is *interrupting capacity (IC)*. This is a breaker's ability to stop the flow of electricity when a high-current fault exists (which is dependent on the amount of energy available from the source). Breakers are made with IC ratings as low as 5,000 AIC and range up to 200,000 AIC. A typical lighting and branch power panel might have a 150A main rated at 65,000 AIC with branch breakers from 15A to 30A rated at 10,000 AIC. Most circuit breakers have two modes of tripping: time (thermal) trip and instant (magnetic) trip. Some breakers also have a third mode called *ground fault* (shunt coil). The first type of trip occurs in the case of mild overloads and works by heating a thermal trip device. The time required for this function is related to the amount of the overload current. For example, a 200% overload may trip the breaker in 100 seconds, while a 600% overload can trip in 2 seconds.

Breakers also operate on high-fault overloads via a magnetic trip device. This function is referred to as an instantaneous trip. The time required for this function is from 1/2 to 1 cycle or .0083 to .0166 seconds. The magnetic trip function generally operates at between 10 and 20 times (1000% to 2000%) the rated current and above. The wide variety of breakers available includes many that are designed to differ from these characteristic tripping curves.

Ground fault tripping breakers are the third type. Most breakers do not have this kind of trip function. Nevertheless, it is a requirement for specific applications that involve potential personal hazards. In ground fault tripping breakers, a sensitive current coil is used to ensure that the amperage going out of the breaker closely equals the current returning through the neutral wire. When these two currents are unbalanced by as little as 5 mA, the coil activates a shunt-tripping device. This process occurs as quickly as 8 milliseconds.

Units of Measure. Circuit breakers are taken off and quantified by the individual unit, EA. They should be listed on the quantity sheet by size, type, AIC rating, and manufacturer.

Material Units. The following items are generally included for panelboard circuit breakers:

* Breaker itself and the lugs for attachment

Additional items that may be required are taken off and estimated separately. Some of the more common items are:

* Panelboard
* Wiring
* Tags and markers

Labor Units. The following tasks are generally included as part of the installation costs for each unit of measure:

* Removing the trim from the panelboard
* Installing the circuit breaker
* Connecting the wires to the circuit breaker
* Installing the trim

Items that may require additional labor-hours are taken off and estimated separately. Some of the more common items are:

* Wire and conduit connections to the panelboard
* Stripping insulating jackets from the wires

Takeoff Procedure. Carefully review the plans and specifications with focus on panelboards schedules. Circuit breakers should be taken off and quantified by amps, number of poles, voltage, and how they are fastened—plug-in or bolt-on. Note on the takeoff the brand and model of panelboard in which the breaker is to be installed. Use the catalog or model number if it is available.

SAFETY SWITCHES

Safety switches are intended for use in general distribution and in branch circuits. They provide an assured means of manually disconnecting a load from its source. The salient feature of a safety switch is that its operating handle is capable of being padlocked in the "OFF" position. This feature protects those working on the equipment from the possibility that someone might inadvertently energize the circuit. For some devices, such as fans, not only is a safety switch required by code, but the switch must be installed within sight of the fan.

Safety switches are designated in three different categories. General-duty safety switches are rated from 30A to 600A at 240V AC. They are available in two-, three-, and four-pole styles, fusible or nonfusible.

Heavy-duty safety switches are rated from 30A to 1200A and 250V AC or DC, or 600V AC. They are available in two-, three-, or four-pole and in either fusible or nonfusible models.

Motor circuit switches rated at 30A to 200A are intended specifically for applications where the switch must be within sight of the motor. They are nonfusible and have either three poles or six poles (for dual-speed motors).

The safety switches previously noted are furnished in a number of different NEMA class enclosures. It is important to price the correct enclosure as cost can vary significantly between enclosure types for the same switch.

Units of Measure. Safety switches are taken off and quantified by the individual unit, EA. They are categorized in the takeoff by size, type, poles, voltage, and enclosure ratings.

Material Units. The following items are generally included per unit:

- One safety switch
- Four lead anchor-type fasteners

Additional items that may be required are taken off and estimated separately. Some of the more common items are:

- Cable and conduit terminations
- Tags and labels
- Brackets or supports (if needed)
- Fuses (if fusible type)

Labor Units. The labor-hours for each safety switch generally include these items:

- Receiving
- Material handling
- Measuring and marking the switch location
- Installing the switch on a masonry wall
- Connecting and phase marking wire on primary and secondary terminals

Items that may require additional labor-hours are taken off and estimated separately. Some of the more common items are:

- Primary or secondary conduit
- Backboards, brackets, or supports for enclosure
- Wire

Takeoff Procedure. Review the plans and specifications carefully for location and types of safety switches. On the quantity sheet, separate the switches by duty type, voltage, number of poles, NEMA classification, and ampere rating. Summarize the total of each for estimating.

Cost Modifications. If several disconnects are being installed in a common location by the same electrician or crew, the following percentages can be deducted from the labor units:

0	to	5	−0%
5		10	−5%
Over 10			−10%

Table 15.5 provides guidance for labor-hours to install safety switches.

Table 15.5   Installation Time in Labor-Hours for Safety Switches

Description	Labor-Hours	Unit
Safety Switch NEMA 1 600V 3P 200 Amps	6.150	EA
NEMA 3R	6.670	EA
NEMA 7	10.000	EA
NEMA 12	6.670	EA

Source: Reprinted from *RSMeans Estimating Handbook,* Third Edition, Wiley.

SWITCHBOARDS

Switchboards are used in buildings that have larger load requirements than can be serviced by a single load center panelboard and its disconnect device. A *switchboard* is a modular assembly of functional sections or compartments. These include the service section, auxiliary section, metering compartment, and distribution sections.

The service section contains the main (incoming) breaker, which may be rated from 200A to 4000A. An auxiliary section is a blank compartment that may be used to facilitate cable pulling. A transition section is provided where needed to interface with transformers, special distribution sections, or an MCC. A metering compartment may contain current transformers (CTs), potential transformers (PTs), and relays, as well as meters for amps, volts, and watts. The distribution sections may contain any combination of fused disconnect switches, branch circuit breakers, or motor starters. The branch circuit breakers may be rated from 30A to 1200A.

Bus bars are available in ratings ranging from 200A through 4000A at 250V DC or 600V AC. Bus bars are generally braced to withstand 50000A fault currents, but optional bracing is available to increase that rating up to as much as 200000A. Bus bars are aluminum, but copper bars are an available option.

Three options are available for wiring: Type A—without terminal blocks; Type B—with terminal blocks for control wiring at the side of each unit; and Type C—with master terminal blocks at the top or bottom of each switchboard section.

Units of Measure. Switchboards are taken off and quantified by section and must be itemized per section, EA; the subcomponents of each distribution section must be listed and detailed per unit, EA. Note any modifications or extras (such as the copper bus option) and quantify accordingly (e.g., per section).

The material and labor units for each section type generally assume that the entire unit is specified in detail during the procurement phase and that the equipment will be factory assembled with all requisite components in place. Some sections may be shipped separately and bolted together in the field.

Material Units. The material cost units for switchboards include:

- The equipment as shipped from the manufacturer
- Bolts, bus ties, and insulators to connect shipping segments
- Anchor bolts and fasteners
- Minor components (such as relays), which may be shipped for field insertion—does not include wiring or connections to such devices by field personnel
- Shipping costs

Additional items that may be required are taken off and estimated separately. Some of the more common items are:

- Equipment pads
- Steel channels
- Special knockouts
- External controls
- Conduit
- Wire and cables

Due to the unique nature of switchboards, they should always be priced by the manufacturer. Pricing should include shipping and any lead times associated with the piece of equipment.

Labor Units. The following procedures are generally included for the installation of switchboard components:

- Receiving and equipment handling to location
- Assembly of shipping sections
- Anchors and bolting in place
- Leveling, shimming, and alignment
- Interconnection of bus bars and wiring
- Marking panel legend

Items that may require additional labor-hours are taken off and estimated separately. Some of the more common items are:

- Conduit
- Wiring and terminations
- Wire tagging and identifications
- Testing
- Adding breakers, switches, and meters to existing switchboards

Takeoff Procedure. Review the plans and specifications for quantity, location, and specifics of switchboards. Switchboards and their associated components should be quoted from the specified equipment manufacturers as a factory-assembled unit. The following items should be included:

- Switchboards
- Feeder sections
- Instruments
- Distribution sections and subcomponents
- Options and modifications

Table 15.6   Installation Time in Labor-Hours for Multi-Section Switchboards

Description	Labor-Hours	Unit
Main Switchboard Section 1200 Amps	18.000	EA
1600 Amps	19.000	EA
2000 Amps	20.000	EA
Main Ground Fault Protector 1200-2000 Amps	2.960	EA
Bus Way Connections 1200 Amps	6.150	EA
1600 Amps	6.670	EA
2000 Amps	8.000	EA
Auxilliary Pull Section	8.000	EA
Distribution Section 1200 Amps	22.220	EA
1600 Amps	24.240	EA
2000 Amps	25.810	EA
Breakers, 1 Pole 60 Amps	1.000	EA
2 Pole 60 Amps	1.150	EA
3 Pole 60 Amps	1.500	EA

Source: Reprinted from *RSMeans Estimating Handbook,* Third Edition, Wiley.

The takeoff procedure must identify the equipment by section and by the subcomponents within each section. Parameters such as amperage, voltage, type, bus material, and breaker sizes, should be detailed. List the sections by amps, volts, and type. Identify the section's components by type and size. List any options or modifications per section and summarize for estimating.

Table 15.6 provides guidance for labor-hours to install multi-section switchboards.

SUBSTATIONS

A *substation*, or secondary unit substation, is an assembly comprised of a high-voltage incoming line section; a step down, three-phase power transformer; and a secondary low-voltage distribution section. Substations have two principle applications, both of which are industrial. The first is as an incoming service center; the second is as a power center with dense, high-power requirements.

There are several distinct advantages in locating the power transformer and its distribution section close to the load. The most important of which is the cable and/or bus lengths can be reduced. There is also better voltage control, reduced power loss, and a minimum installed cost.

Typical transformer ratings are from 112 kVA through 2500 kVA. The most common primary voltages are 4.16 kV and 13.8 kV, but the available range is from 2.4 kV through 34.5 kV. The most common secondary voltage is 480/277V, but 208V, 240V, and 600V are also used.

Depending on the application, the high-voltage line section may be a terminal compartment for cables to connect to the transformer, or it may contain a disconnect switch (fused or nonfused).

The distribution section is an attached switchboard that usually uses draw-out circuit breakers for power distribution. The bus ratings typically run from 800A to 4000A. The draw-out breakers can be operated manually or electrically. Distribution sections with molded case breakers are also available. In most cases, one compartment of the distribution section is devoted to meters, instruments, and trip relays.

Units of Measure. Each substation must be taken off individually and listed in terms of its component sections and their subcomponents in terms of EA.

As is the case with most major electrical assemblies, substations are normally specified for purchase as a factory-assembled unit. The various sections, however, are usually split for shipping and are then set and bolted together in place at the site. Draw-out breakers are shipped individually to be slid in after the equipment is set.

Material Units. The following items are generally included in the material cost unit:

- The transformers
- Incoming line section and switch
- Distribution section with breakers
- Anchor bolts
- Bus bars and bolts for interconnections

Additional items that may be required are taken off and estimated separately. Some of the more common items are:

- Equipment pads and rails
- Rigging
- Conduit
- Cable
- Terminations

Due to the unique nature of substations, they should always be priced by the manufacturer. Pricing should include shipping and any lead times associated with the piece of equipment.

Labor Units. The following procedures are generally included per unit:

- Receiving
- Equipment handling
- Setting, shimming, and leveling
- Anchoring in place

Items that may require additional labor-hours are taken off and estimated separately. Some of the more common items are:

- Cables and terminations
- Potheads
- Testing

Takeoff Procedure. Review the plans and specification for quantity, location, and specifics of substations. List each substation individually by size, voltage, and type of transformer. Then list each component section with its subcomponents. Summarize for estimating.

Table 15.7 **Installation Time in Labor-Hours for Transformers**

Description	Labor-Hours	Unit
Oil Filled 5 KV Primary 277/480 Volt Secondary		
3 Phase 150 KVA	30.770	EA
1000 KVA	76.920	EA
3750 KVA	125.000	EA
Liquid-Filled 5 KV Primary 277/480 Volt		
Secondary 3 Phase 225 KVA	36.360	EA
1000 KVA	76.920	EA
2500 KVA	105.000	EA
Dry 480 Volt Primary 120/208 Volt Secondary		
3 Phase 15 KVA	14.550	EA
112 KVA	23.530	EA
500 KVA	44.440	EA

Source: Reprinted from *RSMeans Estimating Handbook,* Third Edition, Wiley.

Table 15.8 **Installation Time in Labor-Hours for Secondary Unit Substations**

Description	Labor-Hours	Unit
Load Interrupter Switch, 300 KVA and below	60.000	EA
400 KVA and above	63.000	EA
Transformer Section 112 KVA	37.000	EA
300 KVA	59.000	EA
500 KVA	67.000	EA
750 KVA	83.000	EA
Low-Voltage Breakers		
2 Pole, 15 to 60 Amps, Type FA	1.430	EA
3 Pole, 15 to 60 Amps, Type FA	1.510	EA
2 Pole, 125 to 225 Amps, Type KA	2.350	EA
3 Pole, 125 to 225 Amps, Type KA	2.500	EA
2 Pole, 700 and 800 Amps, Type MA	5.330	EA
3 Pole, 700 and 800 Amps, Type MA	6.150	EA

Source: Reprinted from *RSMeans Estimating Handbook,* Third Edition, Wiley.

Table 15.7 provides guidance for labor-hours to install transformers.

Table 15.8 provides guidance for labor-hours to install secondary unit substations.

CONTROL SWITCHES

Control switches provide manual input to the control circuits of relays or magnetic starters. Control switches are used for equipment such as valves, fans, pumps, conveyors, and air conditioners. Control switches are assembled or "built up" using the following three types of components: the legend plate, the operator mechanism, and the contact blocks. Each component is available in a multitude of styles and types that can be combined in nearly endless variations to suit the particular application.

Legend plates are used to describe the functions controlled by a particular control switch. They are ordered either by stock descriptions (stop, start, jog, up, down, etc.) or with unique inscriptions. The legend plate is usually included in the material and labor pricing of the operator.

Operators are the mechanical devices that position the contacts. A whole range of operators is available. Some common types are:

- Selector switch—maintained position
- Selector switch—spring return
- Push button—spring return
- Push button—with illuminated button

Also available is a variety of indicating lights, styled to match the operator. These lights indicate the operating status of the equipment.

Contact blocks provide control and are mounted behind the operator in modular (stackable) fashion. Contacts are described at their "rest" position as either "normally open (NO)" or "normally closed (NC)". A third contact arrangement is a block that has one NO pole and one NC pole. The majority of switches use some combination of these three basic contact blocks. An operator can move the contact to the "held" position to effect control.

Common operator and contact arrangements are assembled and "in stock," while others must be ordered as components and field assembled. Attention must be paid to the voltage and amp ratings of the contacts when ordering.

Control switches may be installed into distribution equipment such as motor control centers for local control. Control switches may also be installed into control station boxes for remote control functions.

Units of Measure. Control switches are taken off and quantified individually as EA. They are listed in the quantity sheet according to type, contact arrangement, and legend function. Each pilot light is counted and listed according to color and voltage.

Material Units. The following items are generally included in the material cost of each control switch:

- Legend plate
- Retaining rings
- Operator
- Contact block
- Wire marker for each contact

Additional items that may be required are taken off and estimated separately. Some of the more common items are:

- Enclosure
- Wire
- Branch conduit

Labor Units. The following procedures are generally included in the labor-hours for control switches:

- Installing control switches in a prepunched enclosure
- Marking control wires
- Testing switch

Items that may require additional labor-hours are taken off and estimated separately. Some of the more common items are:

- Installing wire to control switch
- Termination of control wires
- Cutting or punching mounting hole
- Installing enclosure
- Branch conduit to enclosure

Takeoff Procedure. The biggest misconception on pricing control switches is that the estimator needs to understand all the functions being controlled by the switches. Most control prints contain a detailed legend of control components, including types and catalog numbers. Thus the estimator need not be involved with the operation of the equipment.

Set up the takeoff sheet by the type of control switch. Quantities are best taken off from a schematic control drawing. Be sure to identify those installations that involve manually punching or cutting mounting holes. Transfer the quantities to the work sheet and extend.

Cost Modifications. Most control station enclosures are prepunched. However, if control switches are to be installed in a custom-made enclosure where the holes must be field determined and punched, the labor cost must be increased. Refer to Cutting and Drilling in Chapter 12, "Raceways," or to other standards being used; then choose the hole size required.

For a large concentration of control switches in the same enclosure, the following percentages can be deducted from labor-hours:

1	to	10	−0%
11		25	−20%
26		50	−25%
51		100	−30%
Over 100			−35%

16 | Transformers and Bus Duct

Transformers are devices with two or more coupled windings, with or without a magnetic core. They introduce mutual coupling between circuits and are used to convert a power supply from one voltage to another. *Bus duct*, or *busway*, is a prefabricated unit that contains and protects one or more busses. In this chapter, there are five sections on: transformers, isolating panels, two types of bus duct, and Uninterruptible Power Supply (UPS) systems. Units for measure, material and labor requirements, and a procedure for takeoff follow each description.

TRANSFORMERS

Transformers are used in alternating current (AC) systems to convert from one voltage to another. They cannot be used in direct current (DC) systems. The total energy entering a transformer is equal to the energy leaving the transformer, except for minor losses. Thus, when the voltage is stepped down, the current changes in inverse proportion, specifically it increases. A transformer's capacity can therefore, be designated in terms of the product of the voltage (V) and the amperage (A), that is, as volt-amps (VA) at either side. Larger transformers are often designated in thousand *volt-amps* or *kVA*.

Transformers are used for four basic applications: (1) instrument transformers, (2) control transformers, (3) isolating transformers, and (4) power transformers. Their capacities range from fractional VA to thousands of kVA. The first two are single-phase types only, but the second two may be single- or three-phase designs.

Power transformers may be either dry-type air-cooled or liquid-cooled. Both can be furnished for indoor or outdoor use. Voltage ratings are always described from the *primary* or incoming side to the *secondary* or output side.

Units of Measure. Transformers are taken off and quantified as individual units, each (EA).

Material Units. The following items are generally included in the material unit price:

- Transformer
- Anchor bolts or screws
- Identification tag

Additional items that may be required are taken off and estimated separately. Some of the more common items are:

- Transformer pad, cast-in-place or precast concrete
- Grounding components
- Support platforms, brackets, and the like
- Conduit wire and cables
- Termination plugs

Labor Units. The following procedures are generally included in the labor-hours for transformers:

- Receiving and handling
- Crane or rigging to set in place (if needed)
- Setting or hanging in place
- Anchor bolts

Additional items that may require more labor are taken off and estimated separately. Some of the more common items are:

- Supports, brackets, and the like (fabrication and installation)
- Conduit and connections
- Cables and terminations
- Grounding
- Testing

Typical Job Conditions. The labor units assume approximately 100′ or less from the delivery area of the transformer to its installation location. Adjustment factors will need to be applied for work in congested areas or for material handling that involves especially long distances or unusually difficult access.

Takeoff Procedure. Review and study the plans and specifications carefully for the transformer requirements, quantities, type and location. List the transformers by VA rating and service voltages on the quantity takeoff sheet. Summarize quantities for pricing.

Table 16.1 provides some guidelines for labor-hours associated with the installation of transformers.

ISOLATING PANELS

Isolating panels are used in hospitals, where they serve as added protection—to patients and sensitive monitoring equipment—from the effects of ground potentials. Isolated panels are designed to provide ungrounded service for a variety of hospital

Table 16.1 Installation Time in Labor-Hours for Transformers

Description	Labor-Hours	Unit
Oil Filled 5 KV Primary 277/480 Volt Secondary		
3 Phase 150 KVA	30.770	EA
1000 KVA	76.920	EA
3750 KVA	125.000	EA
Liquid-Filled 5 KV Primary 277/480 Volt		
Secondary 3 Phase 225 KVA	36.360	EA
1000 KVA	76.920	EA
2500 KVA	105.000	EA
Dry 480 Volt Primary 120/208 Volt Secondary		
3 Phase 15 KVA	14.550	EA
112 KVA	23.530	EA
500 KVA	44.440	EA

Source: Reprinted from *RSMeans Estimating Handbook,* Third Edition, Wiley

applications. They supply uninterrupted power in the event of one line to ground fault, while eliminating the danger of electric shock. Should a fault occur, power is maintained to critical life support systems. This kind of protection is particularly important in a hospital environment because of the numerous electrical instruments and appliances with wet systems. An operating room panel typically consists of a back box with stainless steel trim. It contains an isolating transformer, a primary circuit breaker, eight 20A two-pole secondary circuit breakers, and a line isolation monitor.

The critical care area panels are the same as operating room panels, with the addition of eight power receptacles and six grounding jacks.

The X-ray panels are also similar to operating room panels, but they have one 60A, two-pole breaker instead of eight 20A, two-pole breakers. Up to eight outlets can be controlled from this panel.

Units of Measure. Each panel taken off and quantified as an individual unit, EA. Each unit of measure includes breakers and transformers.

Material Units. A complete unit of measure includes panel with all components and trim similar to the materials units for a load center. Anchors for mounting are generally included in the material unit price.

Labor Units. The following procedures should be included in the labor-hours for the installation of each unit:

- Receiving and handling
- Measure and level panel
- Setting in place with wall-mounting hardware

Additional items that may require more labor are taken off and estimated separately. Some of the more common items are:

- Making the connections to primary and secondary
- Testing
- Structural support

Takeoff Procedure. Review plans and specifications in detail for locations, type, size, and quantities of isolated panels. List the panels individually on the quantity sheet carefully defining each component or requirement of the panel. Summarize for estimating.

BUS DUCT

Bus duct provides a flexible distribution system for power in industrial and commercial buildings. The busway itself consists of copper or aluminum bars mounted in a sheet metal enclosure. There are several types available. The most common are feeder type, plug-in type, and weatherproof feeder type. The first two are intended for interior use, while the third is used in exterior applications.

In some styles of bus ducts, *branch taps,* or the access points for connection, can be readily changed to conform to new locations of motors and equipment. This system is widely used in industrial plants where equipment is continually changed to meet new manufacturing conditions.

Feeder-type bus ducts are used primarily in industrial buildings to connect the service entrance to the main switchboard, or for high-capacity feeders to distribution centers, including feeder risers. Feeder types have no access, or taps between the service starting point, or main, and the point of use.

Plug-in bus ducts are used for indoor systems where flexibility is needed for power over a wide area. It is available in ratings from 225A to 3000A in a three-pole, four-wire arrangement. Sections are available in 1′ through 5′, and 10′ lengths. Provisions for plug-in taps are made through outlets located on both sides of the duct.

Weatherproof feeder–type bus ducts perform the same function as feeder-type busways, but they are enclosed in a weatherproof casing for use outdoors or in damp, indoor areas.

Plug-in Units

After the bus or feeder duct has been installed, plug-in modular units are used to tap or branch off from the main bars of the bus duct. There are various types of modules designed for specific branch circuit needs. Some of the most common are:

- Fusible 30A to 400A capacity
- Fusible with contactor 30A to 200A capacity
- Circuit breaker type 30A to 1,600A capacity
- Circuit breaker with contactor 30A to 225A capacity

Bus Duct Fittings

Standard components are available to provide elbows, offsets, tee connections, and crossovers to a bus duct system. Other fittings include reducers for changing the size of the busway, end closers, and expansion fittings. Cable tap boxes are available to provide cable or conduit tap-offs or feeders. They are made in two types:

- End cable tap boxes for attachment at the ends of a plug-in busway run
- Center cable tap boxes for installation at any point in a busway run

Units of Measure. Straight sections of bus duct are taken off and quantified by linear feet (LF). Any partial sections (less than 10′) are taken off and quantified as an individual unit, EA. Fittings are taken off and listed as individual units, EA, with clear descriptions identifying type.

Material Units. The following items are generally included in the LF cost of bus duct straight sections:

- Bus duct (based on 10′ length)
- Hanger every 5 LF
- 2′ of threaded rod per hanger
- One beam type clamp per hanger

Additional items that may be required are taken off and estimated separately. Some of the more common items are:

- Bus duct fittings
- Support brackets and fasteners
- Auxiliary steel for support
- Wall penetrations
- Scaffold rentals

Labor Units. The following tasks are generally included in the labor-hours for straight sections of bus duct:

- Receiving and handling
- Measuring and marking
- Setup of rolling staging
- Installing hangers (beam clamp type)
- Hanging and bolting sections
- Aligning and leveling

Additional items that may require more labor are taken off and estimated separately. Some of the more common items are:

- Bus duct fittings
- Modifications to structure for hanger supports (auxiliary steel)
- Threaded rod in excess of 2′ per hanger
- Welding and fire watch (if required)
- Penetration through walls
- Firesafing of penetrations (if required)

- Installing fittings, 90's, T's, crosses, and so on
- Testing

Takeoff Procedure. Plans and specifications should be reviewed carefully for bus duct locations. Legends on plans should be reviewed for fittings and components and the location of the bus duct. The quantity sheet should identify type, ampere rating, and voltage for each bus duct and its fittings. Quantities are determined by measuring from the source to the point of use or the end, depending on the type of bus duct.

Note: When pricing the material, do not assume that all bus duct sections are 10′ lengths. Individual manufacturers' standard sales units may vary. The estimator should research the specified product.

While most estimating standards base their LF price on the cost of 10′ sections, those that are less than 10′ have a higher cost per LF. Care should be taken to identify nonstandard lengths. Keep these nonstandard lengths separate from the LF totals and price them using manufacturers' quotations.

Carefully review the plans, count the fittings, and identify them by type, ampere rating, and configuration. Count accessories, such as wall flanges, weather stops, enclosures, switchboard stubs, and spring-type hangers. Summarize the quantities for estimating.

Cost Modifications. Labor-hours are generally based on elevations up to 15′. For bus duct installed above the 15′ height, cost modifications to the labor are recommended to allow for the added complexity. Add the following percentages for higher runs:

16′	to	20′ high	+10%
21′		25′	+20%
26′		30′	+25%
31′		35′	+30%
36′		40′	+35%
Over 40′			+40%

Add these percentages to the labor-hours only for the installation of those quantities that exceed the specified height levels.

BUS DUCT/BUSWAY—100A AND LESS

A 100A plug-in busway follows a similar procedure to a larger busway, as described previously in this chapter. It differs in that it has plated round aluminum or copper bars instead of flat bars. A 100A bus duct is typically used for branch power feeders to panel boards or motors, or for small, distributed loads.

30A and 60A plug-in busways have an open bottom into which the terminal plug is inserted. It is available in 4′ or 10′ lengths and in two- and three-wire configurations. The bus can be cut in the field to length, and capped. A *strength beam* is available to aid in mounting and to add rigidity. This kind of busway is designed to distribute power for all types of lighting, but it can also be used for light-duty power.

Like 30A and 60A busway, 20-amp plug-in busway has an open bottom and has similar applications. It is also used for small portable tools and for drop light cords and receptacles.

Units of Measure. A plug-in busway is taken off and quantified by the LF of straight sections. The required fittings are counted and quantified as EA.

Material Units. The following items are generally included with bus duct for 100A applications and less:

- Straight sections of bus
- Connectors and bolts
- One hanger every 5′
- Beam clamp and 2′ threaded rod for each hanger

Additional items that may be required are taken off and estimated separately. Some of the more common items are:

- Fittings and fasteners
- Scaffolding
- Brackets and fasteners
- Auxiliary steel for support

Labor Units. The following tasks are generally included as part of the installation labor for:

- Receiving and handling
- Installing hangers
- Installing bus duct
- Leveling and alignment
- Setup and breakdown of scaffolding

Additional items that may require more labor are taken off and estimated separately. Some of the more common items are:

- Wall penetrations
- Firesafing of penetrations (if required)
- Supports and auxiliary steel
- Welding and fire watch (if needed)
- Fittings

Takeoff Procedure. Plans and specifications should be reviewed carefully for bus duct locations. Legends on plans should be reviewed for fitting and components locations of the bus duct. The quantity sheet should identify type, ampere rating, and voltage for each bus duct and its fittings. Quantities are determined by measuring from the source to the point of use or the end, depending on the type of bus duct.

Carefully review the plans, count the fittings, and identify them by type, ampere rating, and configuration. Count accessories, such as wall flanges, weather stops, enclosures, switchboard stubs, and spring-type hangers. Summarize the quantities for estimating.

Cost modifications for heights over 15′, discussed in the previous section, can be applied to this section as well.

UNINTERRUPTIBLE POWER SUPPLY SYSTEMS

Uninterruptible power supply or *UPS* systems are used to provide power for legally required standby systems. They are also designed to protect computers and provide optional additional coverage for any or all loads in a facility. Detailed information relating to generator sets will be discussed in Chapter 17, "Power Systems and Capacitors."

This section will address the UPS double-conversion technique. A rectifier/charger coverts AC to DC, and a continuously running inverter converts DC back to AC. If the primary incoming AC power has problems, the inverter will draw power from the batteries.

Units of Measure. UPS systems are taken off and quantified as an individual unit, EA.

Material Units. The following items are generally included per unit of measure:

- Self-diagnostic monitoring and control package battery

Additional items that may be required are taken off and estimated separately. Some of the more common items are:

- Extended runtime batteries
- Warranty enhancement plans
- Battery cabinets, battery racks, and remote control panels

Labor Units. The following tasks are generally included as part of the labor-hours for installation of an individual unit:

- Receiving and handling
- Measuring and marking the location
- Setting in place and fastening (if required)

Additional items that may require more labor are taken off and estimated separately. Some of the more common items are:

- Wire and conduit
- Testing of completed system
- Installation of batteries, racks, and remote control panels

Takeoff Procedure. Review plans and specifications carefully for the UPS specification. Material costs are typically obtained from the manufacturer as a package, including shipping. Note all the pertinent options and accessories specified for accurate material costs.

17 | Power Systems and Capacitors

Power systems are standby or emergency generator sets for providing power to essential services during a loss of normal power. Capacitors are used for power factor correction to increase efficiency, thus reducing electrical utility bills. Definitions are provided for each of the components in this chapter. Units for measure, material and labor requirements, and a takeoff procedure are also provided.

CAPACITORS

Power factor (PF) is a measurement of the ratio between true power and apparent power, that is, PF = W/VA. It is a measure of how effectively the power is being used. The ideal PF is 1.0. In industrial and commercial buildings, reactive loads, such as motors, can cause phase shifts between the voltage (V) and the current (A) which, in turn, create undesirable power factors (0.51–0.80). By installing a capacitive load in the system, the PF can be adjusted toward 1.0.

Industrial *capacitors* are metal-clad units that include one (single-phase) or three (three-phase) capacitors and discharge resistors. Industrial capacitors may be connected to the loads of each individual motor or to the main incoming lines. Capacitors are sized in terms of kVAR at a specific voltage and according to the number of phases.

Units of Measure. Capacitors are taken off and quantified by counting as individual units each (EA).

Material Units. The following items are generally included per unit:

* Capacitor
* Fasteners

Additional items that may be required are taken off and estimated separately. Some of the more common items are:

* Cable
* Terminations

Labor Units. The following tasks are generally included per unit of measure:

- Receiving and handling
- Measuring and marking the capacitor location
- Mounting

Additional items that may require more labor are taken off and estimated separately. Some of the more common items are:

- Wire
- Connecting and marking wire terminals
- Conduit
- Testing

Takeoff Procedure. Plans and specifications should be reviewed carefully for both the location and types of capacitors required. Capacitors can be taken off and quantified simultaneously with the motors and controls. To avoid duplication, verify that they are not included in the price of the switchgear. The quantity sheet should specify voltage, phase(s), and kVAR size. List each capacitor according to its size and type; summarize the quantities for estimating.

GENERATOR SET

Generator sets are used as an emergency or standby power source in the event of the loss of normal power. They are rated in terms of their capacity in *kilowatts* or *kW*. Generators are available in single-phase or three-phase and may be specified at 120V through 6600V. Typical units are used at 120V single-phase or 480V three-phase.

The motors may be fueled by gasoline, propane gas, natural gas, and diesel fuel. Gasoline engines have a low initial cost and start well in cold weather. Gaseous fuels require less engine maintenance and are generally more convenient to supply. Diesel units consume far less fuel than gasoline and gaseous units and are generally available in larger capacities, and the motors require less maintenance. Diesel fuel also has a higher flash point than gasoline or gaseous fuels and, thus, is safer to handle.

In general, gas generator sets are available from 10 to 170 kW, and diesel-powered sets range from 15 to 1000 kW. Larger diesel units are available as custom-ordered units. Although some small models may be started with a recoil rope starter, most units are fitted with electric starters and batteries. Control options include local or remote start/stop functions, engine heaters, exhaust mufflers, and annunciators.

Units of Measure. Generators are taken off and quantified by counting as individual units, EA.

Material Units. The following items are generally included per unit of measure for a complete generator set:

- Engine generator
- Control panel
- Battery and charger

- Muffler (small units)
- Day tank for fuel (small units)

Additional items that may be required are taken off and estimated separately. Some of the more common items are:

- Control and accessory items
- Conduit
- Cables
- Transfer switch
- Weather-tight enclosure
- Mounting pad
- Factory trained technician for start-up

Labor Units. The following tasks are generally included per unit:

- Receiving and handling of the equipment
- Setting, leveling, and shimming
- Anchoring (if required)

Additional items that may require more labor are taken off and estimated separately. Some of the more common items are:

- A muffler and exhaust piping (large units)
- Fuel and fuel piping
- Large fuel tank
- Concrete pad; cast-in-place or precast
- Steel channels other than normally supplied by the manufacturer
- Rigging to set generator
- Connecting of wire and cable
- Testing

Takeoff Procedure. Review the plans and specifications carefully for type, quantity, and location of generator sets. The estimator is directed to solicit quotes for a complete generator set from the manufacturer specified. The generator set is usually priced as a stipulated sum for all parts and components to complete the set. The estimator should review the *Related Work* section of the technical specifications for who performs any of the related work. For example, most generators with an external fuel source will require piping to the generator from the fuel source. It may also require that the piping be below grade, which would require excavation and backfill, another task typically outside the electrical contractor's responsibility. Related work and the responsible party should also be addressed on the quantity sheet.

List each generator on a quantity sheet and summarize for estimating. Note all pertinent options and accessories clearly. Also note any special cranes or equipment needed for handling. Determine if any accessories will need to be field assembled.

Many specifications and some manufacturers will require that a trained service representative be present for initial start-up to validate the warranty. This is usually an added cost and may also be part of the project closeout package of documents for the electrical contractor.

AUTOMATIC AND MANUAL TRANSFER SWITCHES

Transfer switches are used to change from the normal source of power to an alternate source, such as a generator set or a battery-powered inverter. There are two basic types of transfer switches: manual and automatic.

A manual transfer switch is a lever- or handle-operated double-throw device with a pole for each line wire. This type of switch is available from 30A to 600A.

Automatic transfer switches are designed to monitor the normal line source and to switch the load electrically to the backup source should the normal source fail. Automatic transfer switches are rated from 30A to 2000A.

A number of extra-cost options and accessories are available for automatic transfer switches. A few of the options are: auto-start relays to signal a generator when normal power fails; sensing relays to prevent transfer until the backup source is at full voltage; status-indicating lights; timed relay transfer to restore the load to normal power after it has been energized for several minutes; time-delay engine-stop relay to keep the generator running unloaded for a few minutes to cool the windings; and test and exercise control switches to permit the generator set to be started periodically and run.

Units of Measure. Transfer switches are taken off and quantified by counting individual units as EA.

Material Units. The following items are generally included per unit of measure:

- Transfer switch, complete in enclosure
- Fasteners for mounting

Additional items that may be required are taken off and estimated separately. Some of the more common items are:

- Optional control functions (i.e., load shedding)
- Conduit and cable
- Terminations

Labor Units. The following procedures are generally included per installation of individual unit of measure:

- Receiving and handling
- Measuring and marking
- Drilling four lead-type anchors, using a hammer drill
- Mounting and leveling

Additional items that may require more labor are taken off and estimated separately. Some of the more common items are:

- Modifications in the enclosure
- Structural supports
- Preparation and terminating the cable to the lugs
- Circuit identification
- Testing

- Plywood backboard
- Painting or lettering
- Conduit and cable

Takeoff Procedure. Review the plans and specifications carefully for the type (auto or manual), voltage, amperage, and number of poles for each transfer switch. List each on a quantity sheet, being careful to define any required options or accessories. Summarize for estimating.

Note: It is best to get a vendor's quote for accurate pricing, particularly if control options are specified. Also, check the generator package, as the switches can often be purchased from the same manufacturer to ensure compatibility. Frequently, the transfer switch can be included as part of the generator "kit" especially on smaller residential units.

18 | Lighting

By far the most elemental portion of electrical construction is lighting. The first commercial use of electricity was for lighting. Today, virtually every building, industrial plant, house, bridge, and roadway sign makes extensive use of lighting fixtures. In building construction, lighting is still the largest single electrical cost center. Lighting accounts for 30% to 40% of the electrical operating costs in commercial buildings. Therefore, taking into consideration the initial cost of the lighting system, the life cycle costs, and the responsibility of meeting energy conservation requirements, lighting systems play an important role.

This chapter contains descriptions of seven different types of electrical lighting, fixtures, and fixture whips. In addition to the description, each component section includes guidelines on units of measure, material and labor requirements, and a general takeoff procedure. Labor-hour guidelines are suggested for the installation of various types of lighting. Where applicable, cost modification factors—for economy of scale or difficult working conditions—are suggested.

INTERIOR LIGHTING FIXTURES

Fixture styles for interior building lighting can be either surface mounted or recessed in a wall or ceiling. Other options are pendant or hanging fixtures. There are, of course, numerous variations of all of these styles.

Not only is lighting a major cost center for material and installation, but it is also a significant factor in the operating costs of a facility. In the past 15 years, much attention has been given to the development of more efficient lamps. Efficiency in fixtures is measured in the number of lumens per watt. The color spectrum of light from various lamps is also a prime factor in judging efficiency. If a bulb could be made that was, for example, 80% efficient at converting watts to lumens, but the light energy was entirely in the ultraviolet region (invisible to human eyes), its useful output would be zero.

Lamps for interior lighting fixtures can be incandescent, fluorescent, halogen, light-emitting diode (LED), or high-intensity discharge (HID). HID lamps include mercury

Takeoff Procedure. The estimator should review the plans carefully with specific attention to the interior light fixture schedule. Specifications should be reviewed for any particular features or accessories relative to the fixtures. Set up the quantity sheet to correspond to the fixture model number and type as identified in the schedule.

Take off one particular section or floor of the building at a time, counting the number of each type of fixture before going on to the next type. It is also advantageous to include on the same quantity sheet any fittings, canopies, or trims associated with a particular lighting fixture.

Quantities of each fixture type should be summarized for estimating. Except for the most basic of models, fixtures should be quoted by a vendor or distributor for the most accurate pricing.

Cost Modifications. Productivity is based on new construction in an unobstructed floor location, using rolling staging to 15′ high. Add the following percentages to labor only for elevated installations:

15′	to	20′ high	+ 10%
21′		25′	+ 20%
26′		30′	+ 30%
31′		35′	+ 40%
36′		40′	+ 50%
41′ and over			+ 60%

For large concentrations of lighting fixtures in the same area, deduct the following percentages from the labor units:

25	to	50 fixtures	−15%
51		75	−20%
76		100	−25%
101 and over			−30%

EXIT AND EMERGENCY LIGHTING

Exit lights are available with mounting arrangements for walls or ceilings and with or without a directional arrow. Explosion-proof enclosures can be obtained for these lights. Exit signs have either incandescent or fluorescent lamps. Six-inch "EXIT" letters are standard.

Battery-operated emergency lights are available with either battery-mounted or remote heads. Several different types of batteries and voltages are available. Emergency lights are usually surface mounted. (A special lamp pack is available for use in fluorescent fixtures; it mounts either in the ballast channel or on top of the fixture.)

Units of Measure. Exit light fixtures are taken off and quantified individually as EA. Emergency lights are also taken off and quantified individually as EA. If the battery is used with remote heads, count each head (EA) and battery unit (EA) and list separately.

Material Units. Exit lights generally include the exit sign with lamps, anchors, and wire nuts as required. Emergency lights generally include the battery unit and remote head complete with lamps, mounting brackets, anchors, and plug with cord or wire nuts.

Labor Units. The following tasks are generally included per unit of measure for installation:

Exit Lights
- Receiving and handling of fixture
- Layout, measuring, and marking
- Installing the fixture and lamps
- Testing

Tasks such as wiring, boxes, or conduit to feed the exit light are considered additional and are taken off and estimated separately.

Emergency Lights
- Receiving and handling of fixture
- Layout and installing mounting bracket
- Connection in outlet box
- Battery connection and testing

Tasks such as wiring, boxes, or conduit to feed the emergency light are considered additional and are taken off and estimated separately.

Takeoff Procedure. Review the plans for locations of the exit and emergency lighting fixtures, with attention to plan legend for the type of fixture: battery pack, remote heads, double-sided exit sign, and so on. Both exit and emergency fixtures are taken off with the interior lighting package. (Refer to the Interior Lighting Fixtures section earlier in this chapter for guidelines.) Summarize totals of each type for estimating.

EXTERIOR FIXTURES

Exterior lighting serves both utilitarian and decorative purposes. The same assortment of lamps used for interiors, previously discussed in this chapter is also available for exterior use. *Low-pressure sodium* or *LPS* lamps are an additional option for exterior fixtures.

LPS lamps generate an almost monochromatic, yellow light. They are, however, highly efficient, putting out over 170 lumens per watt. They require a few minutes to reach full light output, but will restart immediately after interruption of the power source.

Fluorescent lamps have special concerns when used outdoors. They are sensitive to low temperatures and require more starting energy in cold weather.

Ballasts are available for each of the following temperature ranges: above 50°F for indoor applications, above 0°F for outdoor temperature applications, and above –20°F for outdoor applications. Also, the surface temperature of fluorescent lamps has a direct impact on light output (it drops by 35% at 0°F). There are, however, fluorescent lamps specially designed to compensate for this factor.

Most exterior fixtures are either wall or pole mounted, but can be installed in exterior canopies or soffits. Grade-mounted lights along walkways are called *bollards* and are available in a variety of sizes, shapes and designs.

Units of Measure. Each exterior fixture is taken off by counting and quantified as an individual unit, EA.

Material Units. The following items are generally included per unit of measure:

- Fixture
- Lamp(s)
- Wire nuts
- Fasteners or brackets as required

Additional items that may be required are taken off and estimated separately. Some of the more common items for exterior lighting are:

- Conduit and wire
- Excavation and backfill
- Pole
- Pole base: cast-in-place or precast concrete
- Pole arms
- Photocell control and energy management systems

Labor Units. The following tasks or procedures are generally included per unit of measure for installation:

- Receiving and handling of the fixture
- Layout and measuring
- Mounting or installing up to 15′ above grade
- Lamp and fixture assembling
- Wire connections
- Testing and aiming

Additional items that may require more labor for exterior lighting are taken off and estimated separately. Some of the more common items are:

- Conduit, boxes, fittings, and wire
- Pole and arm installation
- Equipment to set poles or bases
- Excavation and backfill
- Pole base: cast-in-place or precast concrete
- Lightning protection
- Photocell control and energy management systems

It should also be noted that the installation of some types of exterior lighting such as parking lot light poles require some type of equipment to set such as a bucket truck or small sign crane. The same is true for the setting of precast concrete pole bases, typically done with a backhoe or excavator. The estimator should review the technical specification section for the exterior lights with special attention on *Related Work* to clarify responsibility for the setting of pole bases and the related excavation, backfill, and compaction.

Takeoff Procedure. The estimator should review the plans carefully with specific attention to the light fixture schedule. Specifications should be reviewed for any particular features or accessories relative to the fixtures. Set up the quantity sheet to correspond to the fixture model number and type as identified in the schedule. Note any addition components or accessories for the fixture for a complete and accurate takeoff. Take off one particular type of fixture at a time, before going on to the next type. Summarize the totals for each type for estimating.

LAMPS

Refer to Interior Lighting Fixtures and Exterior Lighting Fixtures earlier in this chapter for a detailed discussion of the various lamps that are used in most light fixtures. Fixture costs are normally calculated to include lamps; however, this is not always the case when one receives a quote from a vendor. It is strongly suggested that the estimator confirm that lamps are included in the fixture quote.

Units of Measure. Lamps are taken off by counting and are quantified individually by the unit, EA. Large quantities of lamps can be estimated per 100 lamps.

Material Units. The material cost generally includes only the cost of the lamp.

Labor Units. The following procedures are generally included per unit of measure for lamps:

- Receiving and handling of the lamps
- Use of rolling staging
- Opening of the fixture
- Removal of the old lamp
- Installation of the new lamp and testing
- Hauling the old lamp for disposal

Note: The labor units should presume installing new lamps for entire areas—not just one or two at a time.

Takeoff Procedure. List the number of each type of fixture. Then multiply this figure by the number of lamps in each fixture. List the lamps by type and wattage. Total the quantities and add a few extra for breakage.

Note: The technical specifications for interior lighting occasionally require that new lamps are installed just prior to turn over to the owner. This is especially true if the permanent lighting has been used for any length of time during construction. In addition, the specifications may require that a quantity, frequently a percentage of the total, of each type lamp be turned over to the owner at project closeout as spares or *attic stock*. Both of these can be substantial costs for larger buildings and should be researched in the specifications carefully.

TRACK LIGHTING

Track lighting is a versatile system designed to adapt easily as requirements change. Track lights can be swiveled and rotated to aim in any direction. This kind of flexibility is in highest demand and most often used in applications where focused lighting is required.

The basic component of a track light system is the track, which serves as both a raceway and a source of power for the lights. Tracks may be mounted on walls or ceilings, and come in standard lengths of 4', 6', 8', and 12'. They may contain one, two, or three circuits. To connect sections of track, a joiner coupling is required; this coupling must also be one, two, or three circuits. Joiners are available in the following configurations: straight, L's, T's, and cross-connectors. A feed kit is required to connect branch power sources to the track.

Track lighting fixtures are manufactured in a wide range of styles and finishes. Wattages range from 50 to 300. Fixtures can be mounted anywhere along the track simply by twisting the base of the fixture 180 degrees into the track and depressing a locking tab on the fixture itself. Prices for track lighting and its fixtures can range significantly.

Units of Measure. Consult the specifications for the exact type of fixture, lamp, and any trims or accessories that complete the system. Count the track in standard length segments and list as each length segment as EA. Count the fixtures or track heads as separate units and list as EA.

Material Units. The following items are generally priced separately:

- Track (including mounting screws)
- Light fixtures (track heads)

Additional items that may be required are taken off and estimated separately. Some of the more common items for track lighting are:

- Feed kits
- Joiner couplings
- Switches

Labor Units. The following procedures are generally included in the labor-hours for track lighting installation:

- Receiving and handling
- Measuring and marking
- Installing each component
- Aiming and testing

Additional items that may require more labor for track lighting assemblies are taken off and estimated separately. Some of the more common items are:

- Branch conduit
- Wiring to the track system
- Remote switching or dimmers

Takeoff Procedure. The estimator should review the plans carefully with specific attention to the light fixture schedule. Specifications should be reviewed for any particular features or accessories relative to the fixtures. Set up the quantity sheet to correspond to the fixture model number and type as identified in the schedule. Note any addition components or accessories for the fixture for a complete and accurate takeoff. Take off one particular type of fixture at a time, before going on to the next type. Summarize the totals for each type for estimating according to;

- Track—according to size and number of circuits
- Feed kits by number of circuits
- Fixtures—by type, catalog number, and wattage

Measure the track lengths, noting the different lengths of sections, and transfer this information to the takeoff sheets. Count all feed and end kits and transfer these figures to the takeoff sheet. Then count and list all fixtures.

Cost Modifications. Add the following percentages to labor only for elevated installations:

15′	to	20′ high	+10%
21′		25′	+20%
26′		30′	+30%
31′		35′	+40%
36′		40′	+50%
41′ and over			+60%

FIXTURE WHIPS

A fixture whip is a flexible connection consisting of wire in flexible metallic conduit. A fixture whip is used to connect the fixture, such as a troffer fixture for a suspended ceiling, to an outlet box when the fixture is not mounted directly to the box. Fixture whips are usually 4′ to 6′ long and are intended as a labor saving feature.

Units of Measure. Fixture whips are taken off and counted as individual units of EA.

Material Units. The following items are generally included per unit of measure for material:

- Fixture whip complete with two Greenfield connectors and the ends of the wire already stripped
- Wire nuts for one end

Fixture whips can also be made with plug in connectors at each end. This allows fixtures to be connected and disconnected very quickly.

Labor Units. The following procedures are generally included per unit:

- Receiving and handling of whip materials
- Connecting to fixture and outlet box, or connection of fixture to fixture
- Use of rolling stage in unobstructed area up to 15′ high

Takeoff Procedure. Check the fixture schedule for suspended fixtures. Allow one whip per fixture (or one whip per row of fixtures) as shown on the drawings.

Cost Modifications. Productivity is based on new construction in an unobstructed floor location, using rolling staging to 15′ high. Add the following percentages to labor only for elevated installations:

15′	to	20′ high	+ 10%
21′		25′	+ 20%
26′		30′	+ 30%
31′		35′	+ 40%
36′		40′	+ 50%
41′ and over			+ 60%

For large concentrations of lighting fixtures in the same area, deduct the following percentages from the labor units:

25	to	50 fixtures	−15%
51		75	−20%
76		100	−25%
101 and over			−30%

In summary, there is a general note that applies to most light fixtures and that is accurate pricing always comes from the manufacturer's representative or distributor of the fixture. Typically, an electrical contractor will solicit bids from a manufacturer's representative or distributor of the fixture types early in the bidding schedule. The vendor quote will include fixtures, trims, lamps, and any freight costs, as well as a delivery schedule that can be included within the electrical contractor's proposal. The price of light fixtures can vary significantly based on popularity, design trends, availability, and the manufacturer. It is irresponsible to guess at the price of fixtures.

Tables 18.1 through 18.5 provide some guidance on labor-hours required to install general categories of light fixtures. Specific types of fixtures may vary greatly.

Table 18.1   Installation Time in Labor-Hours for Incandescent Lighting

Description	Hours	Unit
Ceiling, Recess-Mounted Alzak Reflector		
150W	1.000	EA
300W	1.190	EA
Surface Mounted Metal Cylinder		
150W	.800	EA
300W	1.000	EA
Opal Glass Drum 10″ 2-60W	1.000	EA
Pendant Mounted Globe 150W	1.000	EA
Vaportight 200W	1.290	EA
Chandelier 24″ Diameter x 42″ High		
6 Candle	1.330	EA

(Continued)

Table 18.1 (*Continued*)

Description	Hours	Unit
Track Light Spotlight 75W PAR Halogen	.500	EA
Wall Washer Quartz 250W	.500	EA
Exterior Wall-Mounted Quartz 500W	1.510	EA
1500W	1.900	EA
Ceiling, Surface-Mounted Vaportight		
100W	2.650	EA
150W	2.950	EA
175W	2.950	EA
250W	2.950	EA
400W	3.350	EA
1000W	4.450	EA

Wire or cable termination of light fixtures are included in the installation of light fixtures.

Source: Reprinted from *RSMean Estimating Handbook,* Third Edition, Wiley.

Table 18.2 Installation Time in Labor-Hours for Fluorescent Lighting

Description	Labor-Hours	Unit
Troffer with Acrylic Lens 4-32W RS 2' x 4'	1.700	EA
2-32W URS 2' x 2'	1.400	EA
Surface Mounted Acrylic Wrap-around Lens		
4-40W RS 16" x 48"	1.500	EA
Industrial Pendant Mounted		
4' Long, 2-32W RS	1.400	EA
8' Long, 2-75W SL	1.820	EA
2-110W HO	2.000	EA
2-215W VHO	2.110	EA
Surface-Mounted Strip, 4' Long, 1-40W RS	.940	EA
8' Long, 1-75W SL	1.190	EA

Source: Reprinted from *RSMean Estimating Handbook,* Third Edition, Wiley.

Table 18.3 Installation Time in Labor-Hours for High-Intensity Lighting

Description	Labor-Hours	Unit
Ceiling Recessed-Mounted Prismatic Lens		
Integral Ballast 2' x 2' HID 150W	2.500	EA
250W	2.500	EA
400W	2.760	EA
Surface-Mounted 250W	2.960	EA
400W	3.330	EA
High Bay Aluminum Reflector 400W	3.480	EA
1000W	4.000	EA

Source: Reprinted from *RSMean Estimating Handbook,* Third Edition, Wiley.

Table 18.4   Installation Time in Labor-Hours for Hazardous Area Lighting

Description	Labor-Hours	Unit
Fixture, Pendant-Mounted, Fluorescent 4' Long		
2-40W RS	3.480	EA
4-40W RS	4.710	EA
Incandescent 200W	2.290	EA
Ceiling Mounted, Incandescent 200W	2.000	EA
Ceiling, HID, Surface-Mounted 100W	2.670	EA
150W	2.960	EA
250W	2.960	EA
400W	3.330	EA
Pendant-Mounted 100W	2.960	EA
150W	3.330	EA
250W	3.330	EA
400W	3.810	EA

Source: Reprinted from *RSMean Estimating Handbook,* Third Edition, Wiley.

Table 18.5   Installation Time in Labor-Hours for Parking Area Lighting

Description	Labor-Hours	Unit
Luminaire HID 100 Watt	2.960	EA
150W	2.960	EA
175W	2.960	EA
250W	3.330	EA
400W	3.640	EA
1000W	4.000	EA
Bracket Arm 1 Arm	1.000	EA
2 Arm	1.000	EA
3 Arm	1.510	EA
4 Arm	1.510	EA
Aluminum Pole 20' High	6.900	EA
30' High	7.690	EA
40' High	10.00	EA
Steel Pole 20' High	7.690	EA
30' High	8.700	EA
40' High	11.77	EA
Fiberglass Pole 20' High	5.000	EA
30' High	5.560	EA
40' High	7.100	EA
Transformer Base	2.670	EA

Source: Reprinted from *RSMean Estimating Handbook,* Third Edition, Wiley.

19 | Electrical Utilities

Electric utilities include the site work required for the installation of electrical power or telephone/data wires and the actual underground installation of those wires or cables. This chapter addresses electrical and communication site work and underground duct bank. Each of these components is described in terms of what is included and required—for both materials and labor. A takeoff procedure is also provided.

ELECTRIC SITE WORK

This category covers the distribution methods used to route power, control, and communications cables onto a facility's property and between its buildings and structures. There are three basic options: direct burial cables, underground in duct banks, and overhead on poles.

Direct burial cables are the least versatile; they are generally used for residential applications where aesthetics are a more important factor than flexibility or an allowance for future changes. Occasionally, direct burial is used in commercial and industrial facilities to route a branch feeder to unique equipment, such as a well pump or roadway lighting. Two techniques are used to place direct buried cables. A trenching machine may dig a narrow slot two to four feet deep for a single cable or a trench one to two feet wide for multiple cables. The trench is usually backfilled with a few inches of sand—below and again above the cables—for protection. About 12 inches above the cable, concrete planks or plastic marker tape may be placed as a warning to future excavation and digging. Finally, on the finished grade, concrete, monuments, or markers maybe placed every 100′ to indicate the path of the buried cable and to denote changes in direction.

Duct banks are groups of two or more underground conduits usually encased in concrete. Although more costly than direct burial systems, duct banks offer three advantages. First, the cables are far better protected from hazards and the elements. Second, groups of several cables can be pulled through each conduit. Finally, new cables may be pulled (or failed cables replaced) quickly and economically to meet future needs.

For duct banks that cover long distances, access must be provided to the run for pulling cables. When only a few conduits are installed, a hand hole meets this requirement. For multiple conduit banks and for large conduits, a structure called a manhole is installed. Hand holes and manholes may also serve to change the direction of a run or to split up a run. Both are usually made of precast concrete or high-density plastics, and manholes are sized large enough for both personnel and pulling equipment. Pulling eyes (steel inserts) are normally cast into the walls of the precast structure opposite the cable entry points. This is done to facilitate the rigging process for cable pulling. Hangers may also be cast into the boxes to carry or support the installed cables.

Duct banks are generally buried, allowing 2′ to 4′ from grade to the top row of conduits; the conduits are separated by plastic spacers and held fast with tie-wires. When the conduit is encased in concrete, it is important to prevent the conduit from "floating" in the pour. A minimum of 2″ of concrete is typically required on all sides.

The most common duct bank conduit material is polyvinyl chloride (PVC) and fiber duct. Galvanized steel may also be used, especially when power cables and instrumentation will be pulled into separate conduits in close proximity to each other. PVC conduit comes in 20′ lengths and can be bent in the field with a heater. Couplings and fittings are affixed using PVC cement. Unless the run is very short (under 100′), the installation of duct conduits of less than 2″ in diameter is very rare.

Poles and overhead routing represent the most conventional method of distributing power and communication cables. Many cables are built and rated for aerial service. Some cables include "strength members" to carry the tensions of the stretched cables. Still other types of cable, such as service drops and telephone lines, will be supported by a steel messenger wire. The hole for a line pole is usually made with an auger machine. After the pole is set, crushed limestone or similar fill is placed and compacted around the pole. In some areas, poles will need lightning rods and grounding. Any special requirements such as grounding should be noted on the plans or in the specifications.

Units of Measure. Multiple quantity designations for units of measure may apply to electrical and communication wiring site work.

- For direct burial trenching, the work is measured per linear foot (LF).
- For duct bank conduits, the work is taken off and quantified per LF.
- For sand backfill or concrete encasement, measurements are made in cubic yards (CY).
- Precast structures such as manholes and hand holes are counted as individual units of each (EA) with specific attention to size, access holes, castings, and so on.
- Fittings, spacers, and elbows are also counted individually, as EA.
- Poles and crossbars are taken off by counting and quantified as individual units, EA.
- Cable is taken off and quantified by the linear foot and converted to units of 100 linear feet (CLF).

Material Units. The following summaries list the material items usually included in the material unit for each component:

Trenching. Includes no units for actual materials. However, for most electrical contractors the trenching machine will represent a rental cost to be included in the estimate. Even if the contractor owns the trenching machine a cost representing ownership, depreciation, and operating expense should be included within the estimate. Most equipment is rented by the day, week, or month depending on the quantity of trenching. Some rental companies may have half-day rates available. For rented equipment, the estimator must consider the portal-to-portal costs of getting the equipment to and from the site. Rental costs do not include the operator or fuel.

Excavation. Excavation, backfill, grading, and compacting for concrete duct banks require larger equipment than a trenching machine. This work is frequently done with a backhoe or excavator. It is typically outside the expertise of the electrical estimator to estimate and the electrical contractor to perform. The same is true for truck-mounted augers used for installing poles. In most circumstances, the excavation as well as the backfill, grading, and compacting is outside the electrical contractor's scope of work, although not exclusively. For many projects where the electrical contractor functions as the prime contractor (or general contractor), the excavation, backfill, and compaction may be well within their scope. The estimator should read the specifications carefully.

Duct banks may be excavated to the width of the concrete placement so that the sides of the excavation can be used to form the sides of the duct bank. This practice, called *earthen forming,* eliminates the need for traditional formwork at the side of the duct bank. It should be noted that not all specifications allow this practice. For excavation in sandy or cohesionless soils where the vertical walls of the excavation are not stable, traditional forming of the sides of the duct bank with wooden forms may be required. Again, formwork may or may not be part of the electrical contractor's scope of work. The estimator is directed to review the *Related Work* section of the technical specifications carefully.

Additional material items that may need to be taken off and quantified are the sand bedding, cast-in-place concrete for the duct bank, conduits, cable and terminations, caution tape, and possibly flowable fill or lean concrete for backfill in the roadways. Some other materials are outlined in the following;

Underground Conduits
- Conduit and spacers for the duct bank
- Fittings, couplings, and elbows
- Vertical risers for pole connections
- Transformer pads
- Field bends of PVC conduits
- Formwork (if required)
- Tie-wire to secure the conduits to the spacers
- Post markers for above grade locating

Precast Structures
- Precast, high–density, or fiberglass manholes and hand holes unit with cover and inserts (if required).

Poles
- Auger holes for the pole
- Compacted fill or lean concrete around the pole
- Equipment for trailering and setting pole

Additional items to be taken off and estimated are cross-arms, guy wire and anchors, and lightning protection.

Cable
- Cable only

Additional items to be taken off and estimated are messenger wire (if not "built in"), terminations, and testing.

Labor Units. The following is a summary of the activities usually included in the labor unit of measure for each component:

Trenching and Excavation
- Operator for trenching equipment or excavator
- Placement and grading of sand bedding for conduits
- Compacting, backfilling, and grading over the trench
- Placement of caution tape
- Formwork and stripping of formwork at duct bank
- Placement and striking off of cast-in-place concrete

Underground Conduits
- Placing and fitting conduits, spacers, and wire ties.
- Field bending of PVC conduits
- Affixing vertical conduits to poles
- Cutting and terminating conduits at transformer pads

Additional procedures to be taken off and estimated are special fittings and elbows.

Manholes and Hand Holes
- Excavation for structure, gravel, or stone bedding
- Equipment for setting the precast unit in place
- Backfill and grading around structure
- Setting frames and covers
- Installation of racks in precast structure
- Grounding of structure

Poles
- Auger excavation
- Setting the pole
- Backfill with specified material
- Compacting the fill
- Rough grading
- Patching pavement (if required)

Additional procedures to be taken off and estimated are cross-arms, guy wire and anchor, and lightning protection.

Cable

- Cable only

Additional procedures to be taken off and estimated are messenger wire, terminations, and testing. Also consider the crew and equipment required for stringing aerial cable.

Takeoff Procedure. Review the plans and specifications carefully for the limits of the scope of work for this section. Take off each of the previously noted tasks and quantify it separately in the takeoff. Finish each category before moving on to the next. Add linear feet to cables in duct banks or direct burial to bring the wire to the final connection point. For pole terminations this may be a considerable length.

Quantify the amount of trenching or excavation by taking off the distance from termination to termination. The depth of the trench should also be noted as this will impact daily output. Whenever possible the estimator should refer to the specifications for soil type (sandy, clay, rock, etc.). This would be found in the geotechnical report in the project manual or on a site drawing that has test borings or test pits. Test borings or pits will have a legend that describes the strata encountered during the boring. The estimator can use the information to adjust daily outputs for trenching or excavation.

Determining the quantity of excavation for a duct bank is a volume calculation: length × width × depth. All dimensions should be in the same units of measure—*feet* when calculating the volume. Mixing feet and inches will produce and erroneous value for volume. For example the excavation for a duct bank that was 100′ long × 6′ wide × 4′ deep would be calculated as follows:

$$100' \times 6' \times 4' = 2,400 \text{ cubic feet}$$

Since excavation is estimated in CY, not cubic feet, the preceding value must be converted to CY:

$$2,400 \text{ cubic feet} \div 27 \text{ cubic feet per CY} = 88.9 \text{ CY} \approx 89 \text{ CY}$$

Concrete for the duct back is also a volume calculation and should be determined using the same methodology. For duct banks with more than two - 4″ conduits the volume of concrete displaced by the conduits may be significant. For example, a duct bank with a cross-sectional area of 6.48 SF (3′-0″ × 2′-2″) housing 8-4″ PVC conduits will have an area of 2.73 SF for the conduits. This represents approximately 42% of the total area of the duct bank—clearly a substantial amount.

When calculating backfill and sand bedding, consider that compacted soil will lose volume. Average compaction factors will vary with soil types and moisture content. *Estimating Handbook,* Third Edition, published by the RSMeans Company, provides compaction and swell factors for various types of soil.

The electrical contractor not skilled or experienced in excavation, backfill, formwork and concrete placement should solicit quotes from experienced, licensed, and

based upon the number of rooms in the system. In this case, a complete system is quantified as EA. This is a parametric type of estimate used more for budgeting during design development than competitive bidding.

Material Units. A complete component with fasteners, box, and connections (wire nuts) is generally included in the material price.

Additional items must be taken off and estimated separately in addition to the preceding totals:

- Cable and wire
- Conduit and connectors

Labor Units. The following tasks are generally included per individual unit, EA:

- Receiving and material handling
- Measuring, leveling, and mounting as required
- Terminations
- Testing

Additional items that may require more labor are taken off and estimated separately. Some of the more common items are:

- Conduit and connectors
- Wire

Takeoff Procedure. Review the plans and specifications carefully for manufacture and model numbers. Quantify all of the different components of the clock system by counting them on the plans. Take off and quantify the linear footage of wire using the methods discussed in Chapter 13, "Conductors and Grounding." All components of the complete system should be kept together on contiguous quantity sheets. Summarize the information for estimating.

Due to the variety of systems and manufacturers, it is strongly recommended that a vendor price quote be obtained for all material components.

Figure 20.1 provides guidance as to the labor-hours for the installation of various components of a clock system.

DETECTION SYSTEMS

Fire alarm systems and burglar alarm (intrusion detection) systems are similar in both their principles of operation and their installation techniques. Nevertheless, they are two separate and distinct systems requiring different components. Rarely, if ever, do they share any hardware or wiring.

Burglar alarms or *security systems* consist of control panels, indicator panels, various types of detection and alarm devices, and switches. The control panel is usually line powered with a battery backup supply. Some systems have a direct connection to the police or a monitoring company, while others have autodial telephone capabilities. Most, however, simply have local control monitors and an annunciator. The sensing or detection devices are various pressure switches, magnetic door switches,

Description	Labor-Hours	Unit
CLOCK SYSTEMS		
Time System Components, master controller	24.20	EA
Program bell	1	EA
Combination clock and speaker	2.500	EA
Frequency Generator	4	EA
Job time automatic stamp recorder		
Minimum	2	EA
Maximum	2	EA
MASTER TIME CLOCK SYSTEM, CLOCKS & BELLS		
20 room	160	EA
50 room	400	EA
Time clock, 100 cards in and out		
1 color	2.500	EA
2 colors	2.500	EA
With three-circuit program device		
Minimum	4	EA
Maximum	4	EA
Metal rack for 25 cards	1.140	EA
Watchman's tour station	1	EA
Annunciator with zone indication	8	EA

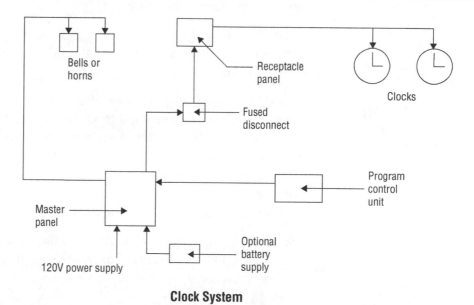

Clock System

Figure 20.1 Installation Time in Labor-hours for Clock Systems

Source: *RSMeans Estimating Handbook,* Third Edition, Wiley.

glass break sensors, infrared beams, infrared sensors, microwave detectors, and ultrasonic motion detectors. The alarms are sirens, horns, and/or flashing lights.

Fire alarm systems consist of control panels, annunciator panels, battery backup with rack and charger, various sensing devices such as smoke and heat detectors, and alarm horn and light signals. Some fire alarm systems are very sophisticated and include speakers, telephone lines, door closer controls, and other components. Some fire alarm systems are connected directly to the fire station. Requirements for fire alarm systems are generally more closely regulated by codes and by local authorities than are intrusion alarm systems.

Units of Measure. The components of each system are taken off individually and quantified as EA.

Material Units. For all of the components identified in the previous paragraphs, the component itself and its mounting hardware are generally included in the material price Additional items that may be required are taken off and estimated separately. Some of the more common items are:

- Wire connections
- Wire
- Conduits

Labor Units. The following items are generally included in the labor-hours per individual unit:

- Receiving and material handling
- Material handling
- Installation and testing

Additional items that may be required are taken off and estimated separately. Some of the more common items are:

- Conduit
- Wire or boxes, unless the box is part of the component
- Termination

Takeoff Procedure. Start by reviewing plans and specifications carefully for the system type, manufacturer, and model of each component. Take off and quantify each component by counting. List them on the takeoff sheet, keeping all components from a particular system together. As previously noted, the takeoff of components should be priced by the manufacturer or authorized distributor. The estimator should compare quantities with the distributor's takeoff, as all components will have a labor component as well.

Takeoff for conduits and wire should follow the procedure in Chapter 12, "Raceways," and Chapter 13, "Conductors and Grounding," respectively. The estimator should consider local codes for wire run in plenum spaces without a raceway.

DOORBELL SYSTEMS

This system includes a low-voltage transformer (120V/24V), push button(s), and a signal device. The signal device can be a bell, buzzer, or chimes. Because the wiring is low voltage, conduit is rarely used.

Units of Measure. Each component is taken off and quantified as an individual unit, EA.

Material Units. The following items are generally included in the material cost per system:

- Doorbell
- Buttons
- Transformer
- Wire terminations

Additional items that may be required are taken off and estimated separately. Some of the more common items are:

- Wire
- Box for transformer

Labor Units. Labor-hours generally include the following procedures for each system:

- Installation of components
- Terminations
- Testing

Additional items that may require more labor are taken off and estimated separately. Some of the more common items are:

- Wire
- Outlet boxes

Takeoff Procedure. Review the plans carefully for the location of the various components. Take off the individual components separately by counting. All components including wire should be listed on the same takeoff sheet to ensure a complete system. Takeoff for conduits and wire should follow the procedure in Chapter 12, "Raceways," and Chapter 13, "Conductors and Grounding," respectively.

Note: The cost of the chime units can vary widely. Be sure to review the specifications when pricing this item. It is always preferable to have a direct quote for chime units from a vendor.

ELECTRIC HEATING

Electric heating is an item that covers a wide range of equipment and applications. A few subcategories are baseboard or wall-unit heaters for residential or commercial buildings, heat trace systems for freeze protection of water lines or for maintaining liquid temperature in chemical and oil pipes, radiant ceiling heat panels, and unit space heaters. Clearly, a number of different estimating problems are posed by an electrical component that covers such a broad range. Nevertheless, two specific applications can serve to guide the estimator through the majority of cases. These are electrical heaters and piping trace cables.

Early in the estimating process, National Electrical Code (NEC) and local building code regulations must be considered as a factor in electric heating. For example, receptacles cannot be located above electric baseboard heaters because of danger to hanging cords. Minimum wall insulation values must also be met when electric heating is installed in residences. It is important to become familiar with these and other requirements before estimating various forms of electric heat.

Electric space heat may be provided with a baseboard arrangement, wall-mounted units, or ceiling-hung heaters. The baseboard units are usually convection types. They are classed in terms of watts per foot (W/LF). Low-watt-density models produce approximately 175 W/LF, while high-watt-density units run 257 W/LF. (The number represents the operating temperature of the element.) Baseboard heaters are available in sizes that increase in 2′ increments—from 2′ to 12′ long. They may be rated for operation on 120V, 240V, or 277V.

Unit space heaters generally operate with forced air (fans) blowing across an electric coil. These units are rated in terms of their heat output in kilowatts (kW). Sizes run from 10 kW to 50 kW. Although 208V to 240V single-phase units are used, 240V and 480V three-phase units are far more common.

There are two types of thermostat controls for heater circuits. The first is full (line) voltage thermostat switches. The second is low-voltage, relay-equipped thermostats. Line thermostats are not suited for more than 240V, nor are they used for large heaters at lower voltages.

If no other information is given about heating capacity requirements, there is a general rule that provides 10 watts per square foot of floor area.

Heat trace is a parallel-resistance heating cable. The purpose of heat trace is to protect a pipe from freezing. This is done by laying a continuous heater wire against a pipe and wrapping both with insulation. Thermostats (with sensing elements against the opposite side of the pipe or at least 90 degrees away) are set to maintain about 40°F. Some liquids, such as heavy fuel oils, require higher temperatures to remain in a fluid state. The heat tracing used on these systems is designed for these higher temperatures. Heat trace systems typically operate from 100° to 300°F. Mineral insulated heat trace cable is commonly used in high-temperature systems.

Heat trace cable is rated in watts per foot, W/LF. If extra heat is needed, the cable can be wrapped in a spiral around the pipe, thereby increasing the number of watts per linear foot of pipe. The number of turns or wraps per foot is called the *pitch*. Tables are available that relate pitch and pipe diameter to total cable footage. Valves, flanges, and other fittings receive extra wrapping to compensate for added heat losses. Special low-temperature heat cable can be used for plastic polyvinyl chloride (PVC) pipes.

When estimating for heat trace systems, it is important to understand the intended use of the cable. Freeze-protection cable is simply attached to the pipe with a polyester tape band every 12″. Another method is to attach the cable to the pipe with a continuous coverage of 2″ wide aluminum tape. This increases heat transfer and temperature distribution around the pipe, allowing higher watt densities without spiral wrapping. Polyester tape bands must still be used every 12″. Cable covered with a factory-extruded heat transfer cement maximizes heat transfer. This covering allows temperature maintenance up to 300°F.

Units of Measure. Heaters are taken off and quantified as individual units, EA. Heat trace cables are quantified in LF.

Material Units. The following items are included for heat component units:

Baseboard Heaters
- The heater
- Mounting screws

Additional items that may be required are taken off and estimated separately. Some of the more common items are terminations, wire, and thermostats.

Unit Heaters
- The heater (with thermostat)
- Brackets or hangers
- Wall or ceiling fasteners as required (up to 15′ high)

Additional items that may be required are taken off and estimated separately. They include terminations, wire, conduit, and thermostats (some models).

Heat Trace
- Heat cable
- Attaching tape

Additional items that may be required are taken off and estimated separately. They include heat cable connection kit, end kit, thermostat, branch circuit wire, and conduit.

Labor Units. The following procedures are generally included in the labor-hours for individual heating units:

Heaters
- Receiving and material handling
- Installation or mounting of heaters in place

Additional items that may require more labor are taken off and estimated separately. Some of the more common items are branch wiring, conduit, terminations, and installation of thermostats (if not integral to the unit).

Heat Trace
- Receiving and material handling
- Installing heat cable by wrapping with tape (up to 15′ high)

Additional items to be taken off and estimated separately are installing end termination kits, terminal kits, thermostats, branch wiring, conduit, and terminations.

Takeoff Procedure. Review the plans and specifications carefully for both the location and the type of heat trace. Set up the quantity sheet by individual components (with heat trace cable, subdivide into volts and watts per linear foot). Add accessories, such as metallic raceway and support clips. Summarize all quantities for estimating.

Notes:

1. If the installation uses transfer cement, note that one gallon applies approximately 60 linear feet of raceway.
2. If the pipe system contains valves, be sure to add enough cable to wrap them. This procedure is described later in this chapter under Cost Modifications.

Cost Modifications for Heat Trace. For each type of valve, add the following quantities to the linear foot totals of heat trace cable:

Butterfly			Flange			Screwed or Welded		
1/2″		0′	1/2″	=	1′	1/2″	=	.5′
3/4	=	0	3/4	=	1.5	3/4	=	.75
1	=	1	1	=	2	1	=	1
1.5	=	1.5.	1.5	=	2.5	1.5	=	1.5
2	=	2	2	=	2.5	2	=	2
2.5	=	2.5	2.5	=	3	2.5	=	2.5
3	=	2.5	3	=	3.5	3	=	3
4	=	3	4	=	4	4	=	4
6	=	3.5	6	=	8	6	=	7
8	=	4	8	=	11	8	=	9.5
10	=	4	10	=	14	10	=	12.5
12	=	5	12	=	16.5	12	=	15
14	=	5.5	14	=	19.5	14	=	18
16	=	6	16	=	23	16	=	21
18	=	6.5	18	=	27	18	=	25.5
20	=	7′	20	=	30	20	=	28.5
24	=	8	24	=	36	24	=	34
30	=	10	30	=	42	30	=-	40

Caution: Pipe insulation is installed after the heat tracing is in place. This is usually performed by the mechanical contractor, but it can sometimes be specified as the responsibility of the electrical contractor. Review the specifications with careful attention to the *Related Work* section.

LIGHTNING PROTECTION

Lightning protection for the rooftops of buildings is achieved by a series of lightning rods or air terminals joined together by either copper or aluminum cable. The cable size is determined by the height of the building. The lightning cable system is connected through a conductor called a *downlead* to a ground rod that is a minimum of 2′ below grade and 1-1/2′ to 3′ out from the foundation wall. To effectively estimate lightning grounding systems, components described in the grounding section of this book will be included. By code, all metal bodies on the roof that are located within 6′ of a lightning conductor must be bonded to the lightning protection system. As a result, the lightning protection system is listed with the equipment grounding in our estimating procedure.

One of the basic components of a lightning grounding system is the air terminal. These are manufactured in either copper or aluminum. The most common sizes used are 3/8″ diameter copper at 10″ high and 1/2″ diameter by 12″ high. For aluminum air terminals, a 1/2″ diameter by 12″ high and a 5/8″ diameter by 12″ high is most common. Air terminals are usually mounted in the perimeter parapet wall and include a masonry cable anchor base. A wide range of configurations are available from manufacturers for mounting air terminals to different roof surfaces.

Cable for main conductor use is calculated in pounds per 1,000′; it is available on 500′ spools. Conductors are manufactured in copper and aluminum. The industry standard Class I minimum weight for copper is 187 lbs. per 1,000′ when used on a roof of less than 75′ high, and Class II minimum weight for copper is 375 lbs. per 1,000′ for structures over 75′ high. The main conductor for Class I aluminum cable is 95 lbs. per 1,000′ for structures under 75′ high, and for Class II, 190 lbs. per 1,000′ for structures over 75′ high. Connections to main grounding conductor cable are accomplished in one of three ways: clamping, heat or exothermic welding, or brazing.

Units of Measure. Air terminals are taken off and quantified as individual units, EA. Cable is taken off and quantified by LF. Terminations are taken off and quantified as individual units, EA.

Material Units. The following items are generally included in the material price of:

Air Terminals
- The terminal or rod
- One mounting base
- Four mounting lead-type anchors and screws

Cable
- Base copper or aluminum cable
- One cable clamp for every three linear feet of cable
- One lead-type anchor and screw every three linear feet of cable

Included in the material cost of ground connections is only the connector itself.

Labor Units. Labor-hours generally include the following procedures:

Air Terminals
- Mounting on concrete surface
- Installing four anchor-type fasteners

Cable
- Receiving main conductor cable on 500′ reels
- Hauling to central roof location

- Setting up cable reel
- Drawing cable
- Fastening cable every 3 linear feet using a lead-type anchor

Connections

- Setting up joint configurations using hardware or molds
- Attaching and terminating the connection

Takeoff Procedure. Review the plans and specification carefully for protection requirements. Set up the quantity takeoff sheet, listing air terminals by size, type of material, and configuration. List the wire size in pounds per 1,000 LF and by type of material. List the connections by type, wire-to-wire size, and configuration. Start the takeoff by counting the air terminals at the perimeter. After the perimeter count, take off all the other air terminals.

Summarize the count of all quantities to the takeoff sheet. Determine the length of the main ground cable, again following the perimeter route. Summarize the quantities, in LF for pricing. Count the number of drops to the grounding electrode, and multiply this figure by the distance between the connection and the ground rod. Measure the equipment cable runs and terminations to the equipment casing and the main grounding conductor. Summarize the quantities to a cost analysis sheet for pricing.

Figure 20.2 provides guidance as to the labor-hours for the installation of various components of a lightning protection system.

Cost Modifications. Labor-hours are based on the installation of up to 10 air terminals on a common roof. Deduct the following percentages from both air terminals and connections for quantities of:

1	to	10	−0%
11		25	−20%
Over 25			−25%

NURSES' CALL SYSTEMS

These systems, used in hospitals and nursing homes, enable the patient to signal the nurses' station, and may also turn on a door light in the corridor outside the room. The central component is a master control station, which includes an amplifier, microphone, and speaker. In some systems, patients can speak to the nurses through a bedside call station or a pillow speaker. It is best to call the manufacturer for a price on the complete system.

Units of Measure. The master control unit, bedside station, and corridor light are all taken off and quantified as individual units, EA.

Material Units. The following items are generally included per unit:

- Master control unit
- Bedside stations
- Corridor light for each room

Description	Labor-Hours	Unit
AIR TERMINALS		
Copper		
3/8" diameter × 10" (to 75' high)	1	EA
1/2" diameter × 12" (over 75' high)	1	EA
Aluminum		
1/2" diameter × 12" (to 75' high)	1	EA
5/8" diameter × 12" (over 75' high)	1	EA
CABLE		
Copper		
220 lb. per thousand ft. (to 75' high)	.025	LF
375 lb. per thousand ft. (over 75' high)	.034	LF
Aluminum		
101 lb. per thousand ft. (to 75' high)	.028	LF
199 lb. per thousand ft. (over 75' high)	.033	LF
ARRESTOR		
175 volt AC to ground	1	EA
50 volt AC to ground	1.190	EA

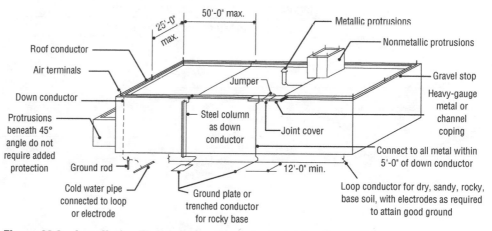

Figure 20.2   Installation Time in Labor-Hours for Lightning Protection

Source: *RSMeans Estimating Handbook,* Third Edition, Wiley.

Labor Units. Generally included in the labor-hours are the following tasks for installation:

- Receiving and material handling
- Installation
- Testing of devices and master control station

Additional items that may require more labor are taken off and estimated separately. Some of the more common items are:

- Conduit
- Wire

- Outlet boxes (unless the box is part of the component)
- Terminations

Takeoff Procedure. Review the plans and specifications carefully for the location of various components and the type. All components are taken off and quantified by individual units, EA. List all the components for the system on the same takeoff sheet(s). Conduit and wire and conduit take off will follow the same procedure introduced in Chapter 12, "Raceways," and Chapter 13, "Conductors and Grounding," respectively. Summarize quantities for estimating.

Figure 20.3 provides guidance as to the labor-hours for the installation of various components of a Nurses' Call system.

Description	Labor-Hours	Unit
Call station, single bedside	1	EA
Call station, double bedside	2	EA
Ceiling speaker station	1	EA
Emergency call station	1	EA
Pillow speaker	1	EA
Duty station	2	EA
Standard call button	1	EA
Lights, corridor, dome or zone indicator	1	
Master control station for 20 stations	24.620	Total

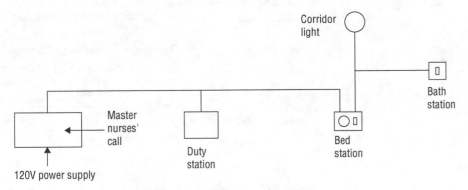

Nurses' Call System

Figure 20.3   Installation Time in Labor-Hours for Nurses' Call Systems

Source: *RSMeans Estimating Handbook,* Third Edition, Wiley.

CLOSED-CIRCUIT TV SYSTEMS

A closed-circuit surveillance system consists of a TV camera and monitor. These systems are used indoors or outdoors for security surveillance. Some applications require pan, tilt, and zoom (PTZ) mechanisms for remote control of the camera.

Educational TV studio equipment is composed of three cameras and a monitor.

Notes: When estimating material costs for TV camera systems and monitors, it is always prudent to obtain manufacturers' quotes for equipment prices. Also, some installations require special accessories. Often, the sales technicians are a good source of information on these requirements.

Some general precautions are:

- Various models of cameras are sold with a wide choice of lens options. Many are priced above "standard" lens costs. It is common to have one type and model camera specified for a site, with each camera having a different lens. As a result, each camera could have a different price.
- Camera models may have several different video tube options. Those designed for "low light" (used outdoors at night) are more expensive than "standard" tubes.
- For long video cable runs, or runs near power cables and equipment, signal accessories may need to be added to the system. These accessories may include hum clampers to remove 60 Hz, AC noise; equalizers to balance the signal's frequency spectrum after a long cable run; and line amplifiers to boost signal strength at the midpoint of the run.
- As the camera-to-monitor distance increases, the type of coaxial cable used becomes critical. High-efficiency cables (low signal loss) are more expensive to buy and to install.
- Monitors are generally specified in terms of resolution in lines. A high-resolution monitor will be more expensive than a lesser resolution.
- Specifications may require the use of expensive test equipment during checkout and start-up. The rental cost or purchase price must be included.
- The presence of the manufacturer's service representative or engineer is often required during testing. This expense will not be included in the material cost and must be added to the estimate.

Units of Measure. Each component in the closed circuit TV system is taken off and quantified by the individual component as EA.

Material Units. The following items are generally included per unit:

- Components, complete with fasteners and brackets

Additional items that may be required are taken off and estimated separately. Some of the more common items are:

- Cable (both 120V power and coaxial signal)
- Conduit
- Terminations (connectors)
- Pole for mounting cameras

Labor Units. The following procedures are generally included per unit:

- Receiving and equipment handling
- Installation and testing of components

Additional items that may require more labor are taken off and estimated separately. Some of the more common items are:

- Conduit
- Wire or cable
- Outlet boxes

Table 20.1 Installation Time in Labor-Hours for Television Systems

Television Equipment	Crew	Daily Output	Labor-Hours	Unit
T.V. Systems not including rough-in wires, cables & conduits				
Master TV antenna system				
VHF&UHF reception & distribution, 12 outlets	1 Elec.	6	1.333	Outlet
30 outlets	↓	10	.800	
100 outlets		13	.615	↓
Amplifier	↓	4	2	EA
Antenna		2	4	"
Closed circuit, surveillance, one station (camera & monitor)	2 Elec.	2.60	6.154	Total
For additional camera stations, add	1 Elec.	2.70	2.963	EA
Industrial quality, one station (camera & monitor)	2 Elec.	2.60	6.154	Total
For additional camera stations, add	1 Elec.	2.70	2.963	EA
For low light, add		2.70	2.963	
For very low light, add		2.70	2.963	
For weatherproof camera station add		1.30	6.154	
For pan and tilt, add		1.30	6.154	
For zoom lens - remote control, add, minimum		2	4	
Maximum	↓	2	4	↓
For automatic iris for low light, add		2	4	

Source: Reprinted from *RSMeans Estimating Handbook,* Third Edition, Wiley.

Takeoff Procedure. Review the plans for location of the various components to the system. Study the specifications for the model and manufacturer of the components. All components are taken off and counted by the individual piece, EA, with the exception of the wire, which is taken off and quantified by LF or CLF.

Summarize the total of each component for estimating. Count the like components and enter each on the takeoff sheet. As previously mentioned, always solicit material pricing from authorized vendors for the specific manufacturer. This ensures contemporaneous pricing of components.

Table 20.1 provides guidance as to the labor-hours for the installation of various components of a television system.

RESIDENTIAL WIRING

For the majority of this text, the discussion has centered around breaking each electrical task or system down into its most element parts for accurate estimating. An alternative method for residential electrical work employs a somewhat different strategy.

Many residential electrical contractors estimate their work by groups of tasks that combine to form a system or assembly. These well defined systems, when combined, comprise all the electrical components needed for a complete residential electrical installation. Because of the relatively lower cost of residential electrical work in comparison to commercial, and the competitive nature of the residential business, most estimating practice combines the expense of assorted materials and the labor to build or install the system as a unit. This approach involves slightly more risk than the unit price method, but can be very accurate especially when repetitive projects of a similar nature form to create a solid historical data base.

So just what is the system approach? Imagine combining a predetermined linear footage of wire, a rough-in box, a single-pole switch, cover plate, terminations, and the labor for these tasks into a single system. While the linear footage of wire may vary from switch to switch, the cost is still reasonably accurate due to the low cost of the wire. A similar system might be the cost of a service from the weather-head to the panel, or the circuit for an electric range complete with receptacle, wire, terminations, and circuit breaker.

For the electrical estimator who may bid multiple residential units in a week or even a day, this strategy can be a tremendous time saver.

This methodology is not without its disadvantages and risk. For many new homes being built today, it may be difficult to create systems or assemblies for all but the most basic type of homes. Many residential contractors are incorporating electrical features routinely in homes that were once reserved for commercial applications. These include energy management, smart-home features, green systems that regulate energy usage, and a wide array of sensors and controls. The estimator is advised to use care when using an assemblies methodology in these types of homes.

Residential wiring can be broken down into the following categories:

- Service
- Branch circuit wiring
- Appliances
- Heating and air conditioning
- Intercom and/or doorbell systems
- Cable and Internet systems
- Entertainment systems
- Light fixtures
- Special needs

For purposes of residential estimating, this text defines *branch circuits* as circuits containing any combination of receptacles, switches, and lighting outlets. Using a system approach, the estimator need only count the number of these items.

Appliance circuits are defined as specific needs, or direct-connected outlets. Examples are range circuits, water heater circuits, exhaust fans, disposal wiring, and dryer circuits. These circuits are run to, but do not include, the appliance itself. There is little variation in the voltage and amperage requirements of appliances. Thus, many estimating standards apply the same system approach to both appliance circuits and branch circuit devices or outlets.

Additional procedures to be listed and estimated for appliance circuits include:

- Thermostat or controls
- Home run conduit or wiring (over 20')

Residential Intercom Systems

- Measuring and marking
- Mounting retaining rings
- Drilling wood studs
- Running low-voltage wire
- Tie-in of speakers and low-voltage transformer
- 110V feed to transformer

Takeoff Procedure. Review the plans and specifications carefully for location and type of system. Set up the work sheet and identify the units by type and defining characteristics.

For example: 115V receptacle SP switch

Count all units of one type before proceeding to the next. Mark all components with a colored pencil. Home runs should be measured back to the panel. Summarize all quantities for estimating.

The fixtures should be taken off last. Manufacturers' prices should be used when appropriate, since fixtures are the single most expensive component and their cost has a significant impact on the estimate. Because of this effect, many contractors issue a fixture allowance for each house or customer, with the electrical contractor responsible for installation.

Note: If using the system approach, the subcomponents of each system can be identified and listed separately, then extended against the unit count. This gives the estimator a quantity count of all subcomponents that need to be purchased.

21

Electrical Demolition and Temporary Facilities

Electrical demolition involves the removal of existing electrical features and components to facilitate the installation of new work. Electrical demolition includes a wide variety of systems and tasks—in fact, just as many as there are new systems to install.

Demolition can include the legal disposal of the debris generated or can assign that to another party. Demolition can include removal and salvage for reinstallation or it can turn the removed items over to another party for reclamation or salvage value.

Demolition can include partial removal, as in the case of removing wires from a conduit but leaving the conduit in place for reuse.

Some demolition can be considered part of the entire process of estimating the task. For example; removing and replacing ballasts in light fixtures. The removal of the old ballast, while not technically demolition, can be estimated as part of the entire scope and not a completely separate process.

Regardless of the specific application, there are some considerations that are applicable to estimating all electrical demolition. First and foremost is understanding the full scope of the demolition, that is, exactly what is being removed and what work is to remain. Second, of equal importance is what is the electrical contractor's responsibility is to the demolished items. While these questions may seem relatively straightforward, in practice they can add a layer of complexity to the estimating process. For example, use of the salvage value from items such as copper wire to offset the cost of the work.

Not understanding what work stays and what is removed can result in reinstalling new work for free for items originally scheduled to remain.

This chapter explores some of the more important considerations for estimating electrical demolition.

DEFINING THE ELECTRICAL DEMOLITION SCOPE

All electrical demolition can roughly be grouped into the same categories as new work—wire, cable, raceways, light fixtures, and so on. However, most electrical demolition is estimated in one category: demolition. As stated previously, defining the exact scope can be challenging, and often subject to interpretation. Electrical estimators are directed to start the clarification of the demolition scope by reading the electrical technical specifications section. It is not uncommon to have language in technical sections of Divisions 26, 27, and 28 outlining the demolition and directing the reader to the *Selective Site Demolition* and *Selective Demolition* sections of *Division 2—Existing Conditions* for more information.

Occasionally, electrical engineers will include as part of the electrical drawings set, a plan identifying what is scheduled for removal and the limitations of that removal, for example, the starting and ending points for demolition of a cable tray system, and the point at which the new system connects to it. Existing conditions drawings are more the exception than the rule. Frequently, demolition may be indicated on the same drawing as new work or only defined by the written word with no actual plans for illustration.

Other options include the use of schedules to identify existing panels or fixtures to be removed. The estimator is directed to review these carefully as well as any notes or language that require a portion of the electrical power and lighting be maintained during the demolition. The practice of keeping existing electrical services in operation while the new work is being installed in not uncommon.

The estimator is directed to carefully review any specifications relative to demolition, whether in Division 26, 27, or 28 or in Division 2. If the scope is still unclear, ask for clarification by way of a Request for Information (RFI).

The second part of the demolition process is what is to happen to the debris generated from the work. Specifications will vary significantly on this matter alone. The three main options are typically:

- *All debris generated from the electrical demolition is dropped to the floor where it occurs. No further action required.* This tends to be the simplest option and can be a result of union jurisdictional issues, or simply that a laborer costs less than an electrician to do clean up.
- *All debris generated by the electrical demolition is disposed of to an onsite dumpster provided by others* (typically the owner or general contractor). This option provides that the electrician is responsible for the debris until it is in the dumpster.
- *All debris generated by the demolition is sole responsibility of the electrical contractor to remove from the site and legally dispose of.* This option transfers the responsibility of the disposal to the electrical contractor with the caveat that the disposal be done legally. This option can add a layer of complexity to the estimate especially if there are hazardous materials within the debris (e.g., ballasts, fluorescent lamps, transformer oil).

The first two options are less clear as to which party has ownership of any salvage value or hazardous material disposal. The last option is less ambiguous but still could provide a challenge in the event of substantial salvage value. It is also not uncommon to see a demolition specification that is mix of all three, with language as to the disposition of specific items demolished—for example: All copper wire #4 or larger will be turned over to the owner and delivered to a predetermined location on site. All salvage value will be retained by the owner.

The estimator should bear in mind that a dispute over demolition debris at the early stages of a project can make for a long project. It is always better to ask for clarification than to make assumptions that prove to be incorrect.

ELECTRICAL DEMOLITION

Electrical demolition is by definition labor intensive. It has few, if any, material costs in comparison to most electrical tasks. It can, however, require equipment and expendables to complete. This combined with the labor intensiveness of the work merits careful analysis when estimating.

As with all electrical tasks, electrical demolition can be both an individual task and a crew task. It is dependent on the task itself and the schedule allowed in which to perform the work. As a first step in estimating electrical demolition, it is helpful to separate tasks into what can be done by an individual and what tasks require a crew.

It should also be noted which demolition items will require equipment and which tasks are done by hand. The same holds true for expendables: saw blades, abrasive cut-off wheels, and so on.

Units of Measure. The units of measure for electrical demolition will vary greatly with the task. For many of the tasks, the units will be the same as the new work, for example, pulling wire into a conduit is quantified by the linear foot (LF). The same is true for removing the wire *from* the conduit. It is also quantified in the takeoff by LF.

For many electrical demolition tasks, the work is not selective, and as such, quantification of the task might be better served by a number of labor-hours. For example, if 100 LF of partition were to be demolished and removed by another contractor, the sole responsibility of the electrical contractor would be to deaden the power to the partition's electrical devices. This process, called *make-safe,* may require only the labor to disconnect the circuits to the partitions and then check the devices for confirmation. The units of measure for this task would be *labor-hours,* or *LH*.

Alternative units of measure might include the compilation of a series of procedures or tasks that result in the demolition. This could be classified as a *lump sum*, or *LS*, for example, the transfer of circuits from a panelboard scheduled for demolition to another temporary panelboard, followed by the removal of the first panelboard. This might be best classified as an entire process with the unit, LS. It still is predominantly labor-hours, but easier to estimate as an entire process.

In summary, there is far more flexibility in the units of measure for estimating electrical demolition than there is for estimating new electrical work.

Material Units. Traditionally, there are no significant materials associated with electrical demolition other than the miscellaneous costs of expendable items: wire nuts or boxes and covers and the cost of disposal.

Labor Units. The following tasks should be included per unit of measure for electrical demolition:

- Identifying and marking the work for removal
- Cutting, capping, or disconnecting of work
- Removal of work to the floor/grade
- Disposal of debris generated (if required)

Additional items that may require more labor are taken off and estimated separately. Some of the more common items are:

- Rerouting of wire to maintain power
- Updating as-built or record drawings
- Relabeling of panels or circuits
- Wiring for temporary services
- Reconnection and testing of systems
- Rigging or crane service for replacement of equipment

Takeoff Procedure. Plans and specifications should be reviewed carefully for both the location and types of electrical demolition work. All demolition work should be listed separately from new work in the takeoff and estimate. The quantity sheet should specify the disposition of the debris as well as any expected salvage value and the intended beneficiary of the salvage value.

One final note on electrical demolition. In much the same manner as with new work, no demolition is done without the coordination of other trades. Failure to coordinate can result in nonproductive or wasted labor-hours. The estimator must consider this when estimating demolition work.

TEMPORARY FACILITIES

It is fairly routine for the installation and maintenance of temporary lighting, power wiring, and devices to be the responsibility of the electrical contractor during construction. In addition, many specifications can include the cost of providing a temporary service to trailers or support structures. Typically, the responsibility for these services and continued maintenance is often assigned in *Division 1- General Requirement; Section 01 50 00; Temporary Facilities and Controls* instead of *Division 26 - Electrical.*

The electrical estimator is directed to review *Division 1- General Requirements* thoroughly for assignment of temporary electrical facilities responsibility. The estimator should also carefully review the technical specification section *Division 26- Electrical* for references to *Related Work* in other sections especially *Division 1- General Requirements.*

While installing, maintaining, and removing temporary electrical facilities are rarely significant costs to a project, they can be drains on material and labor resources if omitted from the estimate.

Units of Measure. The units of measure for electrical temporary facilities will vary greatly with the task. For temporary power and lighting wiring, LF. For individual devices such as receptacles and lights, EA. The estimator should also note that temporary facilities must be removed once they are no longer needed.

Material Units. Materials associated with temporary facilities may often be reusable. Consider a temporary service with mast, panel, wiring, and receptacles, this can often be used on multiple projects with a minimum expense for materials and labor to rejuvenate the service after each use.

The estimator is directed to previous chapters for takeoff and pricing procedures for various types of work used in temporary facilities.

Labor Units. The following tasks should be included per unit of measure for electrical temporary facilities:

* Tie-ing in temporary service to power or lighting panels
* Stringing lights or power wire
* Rough service lamps and light fixtures
* Boxes, covers and devices for power
* Panelboards/load centers and wiring
* Removal of temporary electrical facilities

Takeoff Procedure. Plans and specifications rarely, if ever show temporary services. The electrical estimator is directed to give careful thought to both location and sequencing of the temporary facilities. Occasionally, the cost of the electrical utility application, engineering, or connection fees can be the burden of the electrical contractor. This should be researched carefully as this cost can often be significant.

All temporary facility and service work should be listed separately from demolition or new work in the takeoff and estimate. The quantity sheet should specify the term of the service, and whenever possible the anticipated number of relocations of the temporary power and lighting to arrive at a more accurate estimate.

22 | Contract Modifications

Up to this point, the majority of the text has viewed estimating from the bidding phase. However, estimating does not end once the project has been awarded and the work is under way.

On all projects, changes occur. Some changes have an impact to cost, and as a result those changes require estimating. While the actual mechanics of the estimating process for change orders is the same, there are different dynamics that the estimator must consider.

This chapter will introduce the tangible differences in the dynamics of estimating change orders.

CONTRACT MODIFICATIONS

As with all projects, things change. The individual reasons for the changes are numerous and varied. Changes can occur in project scope, contract amount, or contract time. It can be one, two, or a combination of all three. Changes to the contract are aptly termed *contract modifications* and can be the source of disagreements or even outright battles between parties. Contract modifications by definition are regulated by the contract in the exact same way that the base scope is governed by the contract. There are many misconceptions about contract modifications that are harbored by owners and architects, the most popular of which is that contractors make all their money on changes orders, after underbidding the project. For the majority of contractors, contract modifications are more of a headache than a revenue stream, especially the small ones. Contract modifications fall into two main categories: *change orders* and *construction change directives*. Each has very specific applications that will not be discussed in this text.

The common thread among all of the change mechanisms is proper documentation and correct pricing of the change. It should be noted that not all contract modifications are automatically an added cost; in fact, some can reduce cost or have no impact on cost at all.

Change Orders

The *change order* is the most commonly used contract modification. It is defined as a written document issued by the design professional after the contract has been executed. The change order authorizes a change in contract scope or adjustment of contract sum and/or contract time. It requires that all three signatory parties—the owner, the design professional, and the contractor—execute the change order for it to be valid. A valid change order, when signed by all three parties, implies acceptance of the changed contract scope, and adjustment of contract sum and contract time. As previously noted, all changes in contract scope have an impact on contract sum and contract time.

A change order most often begins life as a *proposal request* from the design professional to the contactor. The proposal request has specific instructions that define the change in scope that allows the contractor to review the request and analyze its impact on the three parameters addressed earlier. The contractor in turn forwards a *proposed change order* to the design professional for review and approval. After review, the design professional can either approve, reject, or amend the proposed change. If agreed upon, the design professional will issue a formal change order to the contractor for signature. The most commonly used change order instrument is the American Institute of Architects (AIA) document G701. Once signed by the contractor, the design professional, and owner must execute the document. *AIA A201™, General Conditions of the Contract for Construction*, Article 7 mandate that the contractor is not required to perform the work without a formal executed change order.

The estimator must evaluate each proposal request for not only the immediate impact of the change on those affected work items but the global impact the change has on the schedule, project overhead, and main office overhead. The change order, once executed, cannot be revisited if the pricing is in error. It can be revisited only if the terms or scope under which the change order was priced and agreed to vary.

Construction Change Directives

The second mechanism for change on a construction project is the *construction change directive*, referred to as the *CCD*. The CCD is initiated by the owner, written and ratified by the design professional, and signed by the contractor. The CCD allows the owner a unilateral right to changes in the work without invalidating the contract. It is most often employed when there is no consensus on the adjustment to the contract sum or contract time that would be required for a change order. Other situations, such as work that is performed as *time and material work* are better suited to the change directive, especially when the final cost of the change in work and adjustment to contract time are not known. It can also be used when the normal change order process of proposal request to formal signed change order may take longer than desired and will delay the project. The signed CCD provides reasonable assurance that a fair and equitable adjustment to the contract sum and contract time will be made. Once in receipt of the executed CCD, the contractor must perform the work. The time and material method requires the contractor to provide a detailed accounting of all of the expenditures related to the change in scope. This accounting, backed up by supporting data, such as material invoices, time cards, and equipment usage, is submitted for review and approval by the design professional. Ultimately, the CCD can be converted to a change order when all parties are in agreement and then added to the Application for Payment. The CCD has the unique characteristic of allowing undisputed portions of the CCD to be paid under normal progress payment schedules. Pricing the CCD is less of an estimating task and more of an accounting function.

Minor Changes and Field Directives

Minor changes in the work that occur regularly can be reasonably expected not to require the use of either the change order or the construction change directive. That does not mean that they go undocumented. Minor changes are defined as those that do not affect contract sum or contract time. They render no net change to the contract. Often, these minor changes can be discussed and resolved on-site. They affect work that has yet to occur or impact the critical path of the schedule. The contractor and the design professional agree on the change and the architect issues a *field directive*. There is no requirement for the owner to authorize the minor change. The field directive documents the change for future reference. Field directives are often handled very loosely and sometimes are even forgotten. In the absence of a formal directive issued by the architect, the project manager may choose to document the change by letter, fax, or even e-mail. Whatever the method, field directives must be in writing. They should address the scope change in detail, the date and time that the change was authorized, or the action directed, and should be included in the project record. The field directive should also state that there is no net change to contract time or contract sum.

Pricing Changes in the Work

Proposal requests often require a detailed accounting of the composition of the proposed change order or the construction change directive. The estimator must consider all of the costs associated with the change. The *AIA A201; General Conditions of the Contract for Construction*, as well as many derivatives of that document, address in detail exactly what expenditures are considered "costs" in preparing a change order. The estimator should review this specific language of the contract, and make sure the change proposal is in compliance. Categories include such production costs as material, labor, tools, equipment, rental costs, and the burden on labor. Other, non-production costs include taxes, transportation or delivery charges, bond premium, permit costs, and added insurance costs. The estimator must consider *all* costs when pricing a change order.

When a change order results in the net difference between additions and credits, the allowance for overhead and profit on the change should be calculated on the basis of the net increase. When the change order results in a net credit, the amount of the credit is determined by the costs that would have been incurred in executing the change by the contractor without decreasing the contractor's overhead or profit. The estimator should also be alert for different percentages for overhead and profit depending on whether the work is self-performed by the contractor's own employees or performed by subcontractors.

The actual mechanism for adding the changes to the contract is less important than the correct pricing of the change. Remember, most contracts do not provide for a "second bite at the apple" if the change order is underpriced when signed.

WHAT MAKES CHANGE ORDERS DIFFERENT

As noted previously, estimating change orders follows the same mechanical process as general estimating: review the scope of the proposal request; perform the takeoff to obtain materials quantities, equipment, and labor tasks; price the material, labor, and equipment needed; add the markup; and submit. However, change orders have two additional considerations that make them different:

- How does the change impact the overall schedule?
- Where does the change order occur in the sequence of the work?

The next sections of this chapter will explore these conditions and how they impact the cost of a change order.

Change Orders and the Schedule

Many contract modifications have no impact on the project schedule. They are easily incorporated within the noncritical tasks that comprise a large portion of the Critical Path Method (CPM) schedule. Others, however, can have a major impact on the schedule. These changes clearly lie on the critical path of the schedule and will add time and ultimately costs. These costs are not always materials, labor, and equipment. Sometimes, they include less obvious costs or costs of an ancillary nature. They are still costs and need to be captured in the change order. For example, consider a change order that extends the critical path of a project 10 days for an added piece of equipment. This piece of equipment prevents the permanent heat from being energized, and so the temporary source of heat must be extended 10 days beyond what was anticipated in the schedule. Clearly, the cost of the added piece of equipment, and the labor to install it are part of the change order, but what about the rental on the temporary heaters and the fuel? What about the overhead costs for the home office for extending the project by 10 days? All of these are considerations.

Many contracts have complex legal contract language that provides the contractor demonstrates how a change order or CCD impacts the critical path in order to qualify for an extension of time. This is especially true on projects that have a *liquidated damages* or financial penalties clause for performance failures or delays. Sometimes, the most demonstrative effort to prove a time extension resulting from a contract change can be subjective.

When pricing change orders, the estimator should remember that they work for the company that will perform the work and as such control how the work will be executed. Scheduling and sequencing of work is part of the contractor's "means and methods". Efforts by parties outside the contractor's immediate team that attempt to mitigate the extension should be reviewed skeptically. It is not uncommon for owners and design professionals to try and convince the contractor that the time extension is unnecessary if the work is performed in a different manner.

Before pricing the change order, the estimator should review all change requests carefully for global impact to the schedule. It is in the best interest of the project to consult with those who will perform the work such as subcontractors, vendors, and the supervisor who will oversee its execution. These viewpoints often go a long way in generating a comprehensive review of the change.

It is a fairly common practice of owners to solicit proposals for change requests and then not act upon them. This occurs frequently when the change has no impact to the critical path of the schedule. The problem then occurs when the owner awards multiple change orders that have been in their possession for months three weeks from Substantial Completion. When the multiple change orders are combined and awarded with a short time left in the project, they tend have a much greater impact on the schedule then if they had been awarded in a more timely manner over the duration of the project. While all contracts provide for the owner to make a unilateral change at any time, they also provide that the owner will extend the time and compensate the contractor for the value of the change, if warranted.

When multiple change orders are awarded simultaneously, the contractor is directed to review how the schedule and the project are impacted by this award before accepting the change order. It may be that the initial price submitted for each change order when priced alone is no longer valid and has been complicated by combining multiple changes. For example, a change proposal that was estimated to be performed during regular business hours now has to be performed during off hours due to the lack of available labor resources.

Timing of Change Orders

When estimating the cost of a change order, the most significant factor is *when* the change occurs. The need for a change order may be discovered or requested in advance of the work being done, or it may appear while the installation of the item in question is in process. The need for a change order may also appear when that particular sequence of work has been completed.

Most professionals can understand that a change order requested, priced, reviewed, and issued prior to that phase of the work being started most likely has the least impact on the schedule. As a result, the pricing of the change represents the best value. For example, consider a change order issued for a dedicated circuit, home run, receptacle, box, and plate issued *before* the rough electrical phase has started. It is easy to imagine that this change order could be incorporated in the regular execution of the work with little or no impact to the schedule or productivity. The cost of the change order would reflect the fact that the work would be done with the other rough electrical work.

This might be true of the same change order issued *during* the rough electrical work process. Again, the change order is being issued in sufficient time so as to maintain the flow of work and have little or no impact on the schedule or productivity. Both conditions illustrated in the example—*before* and *during* the rough electrical work in process—is known as *in-sequence* change order work.

Now consider the same change order issued well after all of the finishes have been completed (i.e., drywall, paint, acoustical ceilings, flooring). The work is no longer done in sequence, but clearly *after* the time has passed to incorporate the work in its normal phasing. In this circumstance, the flooring would have to be protected; a rough-in box would be cut into the drywall; acoustical ceiling tiles would have to be removed; the wire would have to be snaked down the wall; the home run to the panel would have to be fished overhead from room to room, removing and replacing ceiling tiles, until finally arriving at the panelboard. Even the novice can understand that this is clearly a different process, with much different costs.

While the material cost might be the same as in the in-sequence example, one can understand that the change order work would entail more tasks such as cutting in the box and moving ceiling tiles not required during the in-sequence application. This would result in additional labor costs and reduced productivity in performing the work *out of sequence* or after its normal phase has passed.

In general, in-sequence work provides the greatest value to the entity paying for the change order while out-of-sequence work does not.

Now imagine that the electrical contractor priced the change order before the rough electrical work was started with the clear expectation that the change order would be done in sequence. If the owner failed to act on the change order and it was issued just prior to Substantial Completion of the project, one could clearly see that the electrical contractor would be hesitant to accept the change order for the original price. When pricing change orders it is important to identify the time in which the work will need to occur.

Another consideration for out-of-sequence change orders is *rework*. Rework can be broadly defined as the demolition or taking apart of recently installed work to accommodate the change order. This can be extended to tasks such as the reprogramming of addressable fire alarm systems after the addition of another device. While it may not fit the exact definition of rework, it does represent a cost originating from the change.

Remember that if a change order causes the cost impact, then that cost should be part of the change order.

EFFECTS OF OVERTIME

Frequently, change orders, especially those issued near Substantial Completion, cannot be performed during normal business hours and are done after regular hours or on weekends. Work performed during these periods is called *overtime*. Overtime, sometimes referred to as *premium* time, is work that is done by an employee that has already worked his or her normal shift for a single day, typically 8 hours. Weekend premium time is defined as work on Saturday or Sunday after an employee has already worked his or her normal workweek, typically 40 hours. Overtime wages exceed those wages earned during normal shift hours, for most it is either 1½ times or 2 times the normal hourly wage.

The use of long periods of overtime, known as *scheduled overtime*, is counterproductive on any project across all trades. Numerous studies by a wide variety of organizations all reach the same conclusion—productivity decreases with scheduled overtime. Short-term overtime, referred to as *occasional overtime*, however, has little impact on productivity. The exact line between scheduled overtime and occasional overtime is less clear.

Table 22.1 is a table that illustrates the effects of scheduled overtime and the cost impact it has.

In addition to reduced productivity, the employer pays more for that hour of work. This represents a difference between *actual* payroll cost and *effective* payroll cost per hour of overtime work. This is due to the reduced production efficiency with the increase in weekly hours beyond the normal 40 hours. As the total hours worked per week increases on a scheduled basis, productivity is lost due to fatigue, lowered morale, and increased accident rate.

For example; assume electricians are working a 6-day week, 10 hours per day, using the data in Table 22.1 after 4 weeks the production efficiency has been reduced to 80% of the normal production. If we consider the average production rate over 4 weeks, the electrician has produced 87.5% of normal production.

$$.875 \times 60 \text{ hours per week} = 52.5 \text{ hours}$$

Mathematically, that means that our electrician has produced the same amount of work that he or she would accomplish in 52.5 regular hours. Since the employer paid the electrician time-and-a-half or 1½-times the normal base wage the actual payroll cost is:

$$\frac{40 \text{ regular hours} + (20 \text{ overtime hours} \times 1.5)}{60 \text{ hours}} = 1.167$$

Table 22.1   Effects of Overtime

Days per Week	Hours per Day	Production Efficiency					Payroll Cost Factors	
		1 Week	2 Weeks	3 Weeks	4 Weeks	Average 4 Weeks	@1-1/2 Times	@2 Times
5	8	100%	100%	100%	100%	100%	100%	100%
	9	100	100	95	90	96.25	105.6	111.1
	10	100	95	90	85	91.25	110.0	120.0
	11	95	90	75	65	81.25	113.6	127.3
	12	90	85	70	60	76.25	116.7	133.3
6	8	100	100	95	90	96.25	108.3	116.7
	9	100	95	90	85	92.50	113.0	125.9
	10	95	90	85	80	87.50	116.7	133.3
	11	95	85	70	65	78.75	119.7	139.4
	12	90	80	65	60	73.75	122.2	144.4
7	8	100	95	85	75	88.75	114.3	128.6
	9	95	90	80	70	83.75	118.3	136.5
	10	90	85	75	65	78.75	121.4	142.9
	11	85	80	65	60	72.50	124.0	148.1
	12	85	75	60	55	68.75	126.2	152.4

Source: Reprinted from *RSMeans Estimating Handbook,* Third Edition, Wiley.

Based on the 60-hour workweek, the cost per hour will be 116.7% of the normal rate at 40 hours. However, because the effective production (efficiency) for 60 hours is reduced to the equivalent of 52.5 hours, the effective cost of overtime is calculated as follows:

$$\frac{40 \text{ regular hours} + (20 \text{ overtime hours} \times 1.5)}{52.5 \text{ hours}} = 1.33$$

Installed cost will be 133% of the normal rate for labor. Therefore, when estimating overtime, the actual cost per unit of work will be higher than the apparent overtime dollar increase in wage, resulting from the longer workweek. These efficiency calculations hold true for those cost factors determined by hours worked.

23 Project Closeout

The final step in the construction project lifecycle is the *project closeout*. While one does not normally think of estimating as having any relevancy to project closeout, it does. Planning for closeout starts in the bidding phase by correctly estimating the cost of various parts of project closeout.

The information and documentation needed for closeout is collected from the very early stages beginning even before the execution of the work. The physical construction work is clearly the largest phase of the construction process however it is not the only phase. There is administrative and contractual paperwork that plays a large part in the transferring of the project from the construction team to the owner's team. Project closeout is not a complicated process, but it can consume time and money.

THE CLOSEOUT PROCESS

The project closeout process begins in the preconstruction phase of the project. It is initiated with a closeout meeting that lays out the contractual requirements with the subcontractors, vendors, and all participants in the process. While it may seem premature, it is quite the opposite. This early introduction to the project closeout process allows the team to develop a system for collecting the necessary documents and information during the execution rather at the very end of the project.

A helpful start to the project closeout process is a careful study of *Section 01 70 00, Execution and Closeout Requirements*, in Division 1—General Requirements of the technical specifications. Many subcontractors believe project closeout is the sole responsibility of the prime or general contractor. Nothing could be more incorrect. Due to the flow-down or pass-through nature of Division 1, all contractors, prime and sub, have a duty to provide closeout documentation. Section 01 70 00 provides a detailed checklist of the items required for closeout compliance for each individual project. The section includes, but is not limited to:

- Closeout procedures
- Final cleaning
- Starting and testing of systems

- Demonstration and instruction to owner personnel
- Protection of installed construction
- Punchlist
- Warranties
- Certificate of Substantial Completion
- Project change order log
- Warranty items in process of correction
- Project record documents; final inspections, as-builts, material test certificates, reports, and similar documents
- Closeout submittals
- Lien releases/waivers
- Operations and maintenance data
- Preventative maintenance instructions
- Maintenance contracts
- Manual for equipment and systems
- Spare parts and maintenance products
- Testing and commissioning reports
- Product warranties and product bonds
- Surety bond releases
- Completed operations insurance policies
- Final site survey (if applicable)
- Extra (attic) stock materials: lamps, devices, fixtures, trims, acoustic tile, and so on
- Rebates and energy credits

While not all of the preceding items are the responsibility of the electrical contractor, many are and, as such, will have a cost associated with collecting and providing the documents and items.

Once the electrical estimator is aware of the general requirements outlined in section 01 70 00, the next step is to create a matrix of what is required for closeout in each of the technical specifications sections for the electrical contract. This can often be found in the Part 1—General paragraphs of each technical specification section. It is recommended that the project team members review Part 2—Products and Part 3—Execution paragraphs as well, to ensure that nothing has been omitted.

The matrix should include a description of the closeout item and provide a space for estimating the cost of the material and labor involved. It should also note who will be responsible for the task so that the proper rate can be applied. Later, after award, the electrical contractor can use the same matrix to assign responsibility for each submission, with the date requested, dated received, and disposition of the closeout item and the date forwarded to the owner/architect/general contractor and the name of recipient. It can actually become a working document for the project manager to track performance.

The electrical contractor's team member heading up the project closeout should always retain copies of all closeout submissions for archiving, as well as signatures acknowledging the receipt. While this may seem a bit drastic, more than one original document has been misplaced by the owner's team and had to be replaced by the contractor or subcontractor at substantial expense and effort because there was no proof of transfer. On the opposite side of the table, this prevents the dispute over what was received and what has not been received.

OBJECTIVES OF THE CLOSEOUT PROCESS

The objective of the closeout process is to create an orderly and documented transition of the project, from possession by the construction team to possession by the owner or end user.

To achieve a successful closeout, there needs to be acknowledgment that the closeout process means different things to different parties. To the owner it can often be a time of great anxiety and apprehension as the operation of the facility becomes their responsibility. Creating a smooth and orderly transition can often cement an already beneficial business relationship, while a hasty exit can leave a lasting impression of abandonment. It is not uncommon for an owner to respond to the abandonment by holding on to the last vestiges of the contract—the contract retainage.

Architects and engineers with construction administration duties have often exercised the owner's contractual rights when the contractor has failed to focus on the completion of the punchlist and closeout procedures in a timely manner. These rights have included:

- The withholding of payment for failing to submit closeout documentation
- Refusing the release of retainage until all closeout documentation has been reviewed and accepted
- Assessing the prime contractor for the Architectural/Engineering firm's time and expenses for prematurely requesting substantial completion or for extending the project closeout procedures ad infinitum.

All of the preceding can be counterproductive to a smooth transition and successful completion of the project. Project closeout procedures can be separated into two main categories: contract closeout and administrative closeout.

CONTRACT CLOSEOUT

Contract closeout involves the fulfillment of the obligations under the agreement(s) that has governed the project. This includes both the prime contract with the owner (or general contractor) and lower-tier contracts with subcontractors. It is also common to verify that the maintenance or service agreements that were part of the original contract are in order and scheduled. Contract closure is a simple, but essential process to project closeout. It often involves a careful review of the governing contract to reconcile items such as allowances and unit prices that were part of the agreement. It should include a careful review of approved change orders to ensure contract sums are correct. This is also done as part of the administrative closeout. The duplication of this task in two different processes acts as a check and balance against error. Contract closeout may also include the matching of purchase orders to invoices to reconcile vendor payments. Any termination of vendors or subcontractors during the construction process should contain an explanation (reasons) of the termination and a reference to the supporting documentation in the event any actions arise from the termination. It should be noted that what seems like a sufficiently detailed explanation at the time of closeout may be woefully inadequate months or years later. A narrative is a method for ensuring that crucial details are not lost. Most of these tasks are performed by the electrical contractor's in-house team: project manager, administrative assistant, and bookkeeping or accounts payables personnel. The estimator must assign a reasonable amount of hours to perform these duties.

ADMINISTRATIVE CLOSEOUT

The administrative closure procedure focuses on the collection, preparation, and archiving of administrative-type documents. Administrative documents are best classified as the financial accounting summary: payment histories to ensure subs/ vendors/suppliers have been paid, regular updating of the closeout matrix, and the archiving of lower-tier documents. It includes the creation and distribution of lessons-learned documentation.

It includes the financial reconciliation of the project and the review of accounts, both internal and external, to verify that they have been closed for the project. Approved change orders are tabulated to verify contract sums and payments made. Disputed amounts that have yet to be resolved are held in abeyance for later disposition. The final job cost report is produced and distributed for review. For many contractors, the cost of the administrative closeout is included as part of the indirect overhead expense and is covered by the home office overhead percentage added in the recapitulation of the estimate.

It would be naive to not acknowledge the some projects end with unresolved issues that may require the involvement of third-party professionals: lawyers, arbitrators, and judges. For projects with unresolved claims, that may mean "going legal." This is a difficult if not impossible item to estimate. A general rule is that if the electrical contractor is trying the estimate to cost of a potential lawsuit, it might be better if the contractor did not bid the project. This is very different, however, than adding a line item in the project overhead for a reasonable fee for the company attorney to review the contract after award.

LESSONS LEARNED

One of the most commonly overlooked lessons to come from a completed project is the cost comparison analysis. The actual cost and how they were expended as compared to the estimated cost at the time of bid. The same is true about schedule. Both are often archived without the benefit of comparing the final costs to the estimated costs or final tasks duration to the estimated duration. Both offer key learning results that have the potential to improve future performance in either category. Sometimes this comparative analysis goes beyond the normal lessons-learned meeting and may be best served by limiting it to the estimating and scheduling departments. This may also include the circumstances that contributed to the difference, if any. The project team must remember that not all lessons learned are tactical; some are simply arithmetical. It would be silly to continue to estimate a task at $5.00 per square foot for labor if there were historical data that supported an actual labor cost of $5.50 per square foot. Use the data as a source of improving future estimating performance.

In addition to the arithmetical lessons learned in estimating and scheduling there are the performance lessons that have been learned. These are crucial especially for the electrical contractor that is growing a business and anticipates specialization in a particular market. Repeating the same mistakes, project after project can be demoralizing at best and costly at worst, yet without the sharing of information that comes from a formal lessons learned process it is hard to avoid.

For the growing contractor with less experienced staff it is a solid foundation for improvement. It also allows the contractor to evaluate policies and procedures for their practical value after the implementation. Lessons learned is the discussion and documentation of the experience gained on the project. Lessons are learned from working with or solving real-life problems that projects often develop. It is the review of the positive and negative situations and how the project team dealt with them throughout the life cycle of the project. It is intended to provide insight for future projects so that the positive experiences are replicated and improved and the negative experiences are mitigated.

Lessons learned meetings are typically conducted like any other meeting. An agenda is written and distributed in advance for any additions by team members. Key stakeholders, the project manager, superintendent, senior executive managers, financial personnel, and in some cases stakeholders outside the immediate team are in attendance. The meeting should be chaired by someone other than the project manager responsible for the execution of the project. This prevents a one-sided view of the project or glossing over of problem performance.

For the lessons-learned meeting to have value the discussion must be frank and honest. Participants must be able to openly present their comments; positive or negative without fear of reprisals. Lesson learned sessions are not a finger-pointing opportunity nor are they intended to be punitive.

Responsibility for problem performance is essential to developing useful recommendations for future projects of a similar nature. Lessons learned meetings are intended to focus on major issues both bad and good. Participants should avoid personal attacks.

Remember to recognize superior performance by members of the project team. Positive reinforcement can be extremely satisfying and rewarding to team members. Team members are generally proud to share positive public recognition by senior management when the project is successful. It can be an enormous motivator for future performance.

Lessons learned meetings can also be the best opportunity to discuss new techniques that were tried, products or processes used or first time subcontractors. These are critical bits of information that will help improve the second use of a product or subcontractor, or more importantly prevent a repeat of a bad experience. Lessons learned sessions are valuable closure mechanisms for team members.

24 | Computerized Estimating

Computers have clearly added speed, power, and accuracy to construction estimating. They make it possible to produce more estimates in less time, break a job down to a more detailed level for better cost control, manage change orders more easily, test "what if" bidding strategies, and integrate estimating with other commonly used construction applications. However, nowhere in the industry has the efficiency of computers simplified and improved the construction process as much as it has for estimating. Automating the estimating function was revolutionary.

The ability to accelerate what once were slow, tedious calculations, combined with the accuracy offered by computers, has allowed estimators to perform these basic functions more efficiently and cost effectively. Leaving number crunching to the computer allows more time for strategizing and exploring new methods to obtain work.

The estimating and bidding process has always been plagued with last-minute changes. Bids submitted late from suppliers and subcontractors to prevent "bid shopping" have always proved problematic. How can last-minute changes be made accurately? Computer estimating builds in the flexibility of making changes—even at the last minute—without having to retrace steps or redo calculations. By changing just one number, the entire estimate can be recalculated automatically. Prior to computers, this process often took a multiperson staff, or at the very least, extra time to check the accuracy of the calculations.

This can now be done in a matter of minutes. Even the cost of performance and payment bonds can be calculated by using a computer-generated algorithm.

The evolution of computers in construction has not been without some minor problems, however. Remember, a computer is a tool to increase accuracy and productivity. In that respect, it is no different than an electric drill or a pipe bender and will perform only as well as the level of expertise of the individual operating it allows. In short, good tools don't make good tradespeople. The same applies to estimating. A computer will enhance your estimating ability but cannot replace it.

There are many levels of computerized estimating, which vary in functionality, sophistication, time required to learn, and, most of all, price. Many estimators make the mistake of immediately transitioning their manual estimating into a fully integrated estimating software system without learning the basics of what a computer can do for the estimating process. Successful implementation of an estimating software system will not happen overnight. It usually takes training and user interaction to get a system working to its full capability.

It is recommended that computerized estimating be introduced through a multistep process. First, learn the basics of computers and estimating through the use of industry-standard spreadsheet software programs, such as Microsoft Excel.

The real value of any computerized estimating system is its cost database. Home-grown spreadsheet applications are often devoid of a database and are used for computation only. The estimator has to plug in each cost—material, labor, and equipment—along with the appropriate formulas to arrive at a number. For smaller contractors, this may be adequate. For larger companies, this may be inefficient and inadequate.

No matter how sophisticated or user-friendly construction estimating software is, its overall success depends on the completeness, functionality, and accuracy of the cost data, and the methods by which it is used. Without a fully functional database, construction estimating software is nothing more than a very expensive calculator.

BASIC SPREADSHEET PROGRAMS

A computerized estimating system should perform three basic functions:

1. Calculate costs.
2. Store and manage data.
3. Generate hard copy reports.

Most industry-standard spreadsheet software programs, such as Microsoft® Excel, meet these requirements. The premise of every spreadsheet program is to automate a calculation sheet. Spreadsheet programs calculate costs in vertical columns and horizontal rows using basic mathematical functions. Table 24.1 illustrates a spreadsheet that might be used by an electrical contractor for estimating purposes.

Another major capability of spreadsheet programs is managing and sorting information. A typical line item of cost data consists of some sort of a line number, description of task, material costs, labor costs, and equipment costs. There is information on crews and productivities that needs to be maintained.

Typical construction databases are in the thousands and tens of thousands of line items. Most spreadsheet programs allow information to be entered, managed, and sorted numerically or alphabetically.

There are some limitations in using spreadsheets. Information within a spreadsheet must be organized using the same field layout. A significant problem is likely to be difficult locating certain line items within the database and bringing that information back into the estimate. Scrolling through a database of thousands of line items

Table 24.1 Sample Spreadsheet Estimate

Bid Date:	July 30, 2014	**Project:**	Renovated Office Building			**Estimate for:**	Base Bid
Time:	2:00 PM		10 Washington St., Boston, MA			**Addenda:**	#1, #2 have been acknowledged
Estimator:	Jack Smith	**Project No.**	2014-23A			**Checked By:**	Mike Jones

SECTION	DESCRIPTION	QUANTITY	UNIT	MATERIAL		LABOR			EQUIPMENT		TOTAL	Check
				UNIT COST	TOTAL	UNIT COST	TOTAL	Labor Hrs.	UNIT COST	TOTAL		
26 51 00	**Interior Lighting**											
1.01	Incandescent, high hat can, round alzek reflector, 100W	56	EA	$70.00	3,920.00	$52.50	2,940.00	49.73	$0.00	0.00	$6,860.00	
1.02	Metal halide, 2'W × 2' L, 250W, integral ballast, recess	33	EA	$334.55	11,040.15	$152.50	5,032.50	85.12	$12.30	405.90	$16,478.55	
1.03	HP sodium, 2'W × 2' L, 150W, integral ballast sur. mtd.	12	EA	$463.20	5,558.40	$155.50	1,866.00	31.56	$14.50	174.00	$7,598.40	
1.04	Wall mounted, metal cylinder, 75W, painted finish	16	EA	$55.00	880.00	$43.00	688.00	11.64	$0.00	0.00	$1,568.00	
1.05	Vaportight, incandescent, ceiling mounted, 200W	23	EA	$81.30	1,869.90	$67.55	1,553.65	26.28	$0.00	0.00	$3,423.55	
1.06	Low bay aluminum reflect, 250W	33	EA	$363.00	11,979.00	$131.25	4,331.25	73.26	$0.00	0.00	$16,310.25	
1.07	2' × 4' fluor., drop in, acrylic lens, 4-32W, T8 lamp	66	SF	$66.50	4,389.00	$88.40	5,834.40	98.69	$0.00	0.00	$10,223.40	
	Interior Lighting Totals				**$ 39,636.45**		**$ 22,245.80**	**376.28**		**$ 579.90**	**$ 62,462.15**	62462.15

is a time-consuming and cumbersome process. Keeping such a database up to date also requires a major investment of time.

The update and maintenance of the database is reason alone why so many electrical contractors, and contractors in general used "canned software" or software with a prepackaged database. The database is maintained and updated, new items and tasks are added, obsolete tasks are culled, and in general the database is contemporaneous. This is a full-time job in and of itself, and often beyond the skills of the average contractor.

While there is no one-size-fits-all application in software, most of the prepackaged software is flexible and can be semicustomized. Many software programs currently available employ the use of a spreadsheet application such as Microsoft® Excel to display the information coupled with the database of the software. Comprehensive as they may be, remember that databases may not represent the costs incurred by *your* firm. This allows for greater flexibility in displaying the information. These are sometimes called *add-on programs*.

Spreadsheet add-on programs are relatively inexpensive, and usually offer regular updates so long as the subscription is active. They generally have a shorter learning curve than stand-alone estimating software.

STAND-ALONE ESTIMATING SOFTWARE

A complete stand-alone estimating system does not require any other software in order to function. It incorporates all of the capabilities previously covered, links to electronic quantity takeoff devices, and the ability to "communicate" with other construction software applications. By integrating with the scheduling module, estimate labor-hours can be used to create and maintain the project schedule. A link to the accounting module allows job cost accounting. Computer-aided design (CAD) and digitizer programs can also be connected to provide electronic takeoff to estimate. Other features include the ability to customize labor rates and markups, input the company's own material prices, and build assemblies, as well as great flexibility in report formats.

Some estimating systems that fall into this category are developed by companies such as RSMeans, Timberline Software Corporation, MC², Estimating Systems, Inc., G2 Estimator, or Building Systems Design. There are hundreds of these systems, and selecting the appropriate one can be a tedious process that involves collecting printed information, seeing and using demos, and evaluating products and their costs.

Estimating software for the construction industry is as varied as the firms that use them. Selecting and purchasing the right software for your company or application can be a daunting task.

Selecting the Best Software

It is important that the estimating software fits the intended use. Selecting a complex system that exceeds the company's needs can prevent it from being used to its full capacity, and can be costly to the company by requiring support services and training. However, purchasing software too simple for the application can relegate it to the dust heap next to the thermal paper fax machines. While these are at extreme opposite ends of the spectrum, they occur with more frequency than one would expect.

Here are some simple suggestions when shopping for estimating software:

- Evaluate the application needs to see if a simple spreadsheet program, such as Excel will suffice or whether a more complex system is required.
- Anticipate that the software should meet approximately 80% of the company's application needs. No "canned" software packages will satisfy them 100%.
- View demonstrations on as many different packages as practical. Ask if the software comes with a demonstration or trial period to ensure that it meets the company's needs.
- Request references from contractors in the same market or type of work who currently use the particular software for additional verification of its usefulness.
- The selected software should allow for some growth. Although a difficult parameter to judge, growth depends on the business cycle of the company. Companies whose sales volume is expanding at an exponential rate may be unable to satisfy this requirement, as they will outgrow software products rapidly as a result of constantly changing needs.
- Make sure the software fits the budget and is cost effective.
- Train personnel who will use the software, multiple personnel if possible.
- Investigate non-trade-specific software first that may suit your needs before looking into programs designed specifically for electrical contractors.
- Assess whether systems can be integrated with other departments, such as job cost reporting, general ledger accounting, or payroll.

While much of the preceding appears to be common sense, it sometimes gets overlooked in the search for ways to increase productivity. Remember, estimating software is not a panacea. It cannot cure bad estimating practices. Regardless of the advertising claims, it will not instantly increase the volume of work. It may, however, eventually allow the estimator to bid more, thereby increasing opportunities.

Do not forget that becoming fluent with any software takes time and is an investment in the company.

QUANTITY TAKEOFF SOFTWARE

In the past 10 years, there have been enormous advances in software for performing takeoff. In the past, takeoff by computer required a digitizer (wand) attached to a computer, some expensive hardware and software, and a lot of training. Takeoff was done by attaching the paper plans to an integrated work surface and a combination of point-and-click tasks with some data entry. New advances in data file storage capacity, the Web, and software no longer requires all the original components. In fact, takeoff can now be done on a laptop while traveling at 30,000 feet in an airliner, with the only requirement being the appropriate takeoff software.

Many architects, engineers, and owners are providing bid documents as electronic files for general contractors and their subcontractors, to use in bidding projects. For quite a few contractors, the plans are never even printed unless they are awarded the project! The architect posts the plans to a File Transfer Protocol (ftp) site. The contractor logs on with the Architect/Engineer's permission, downloads the files, and performs the takeoff and resultant calculations to arrive at the quantities necessary. The estimator then transfers the quantities to the estimating program to generate a bid. Some takeoff programs are actually integrated with the estimating program so that the takeoff becomes the estimate once prices have been applied.

Most programs are available with a free trial period to allow the estimator to see if it will meet his or her needs. Programs without the free trial may not be worth the investment. A general bit of advice: if you cannot invest the time to learn the software, do not invest the money to purchase it.

In summary, computers have a tremendous value in construction applications and they are here to stay. Estimating by computer provides flexibility, accuracy, and speed in producing bids. Its value is limited only by the construction and estimating knowledge—as well as computer skills—of its users. Those looking to purchase computer estimating/takeoff programs should do so only after careful review and understanding of the company's need.

Appendix: Symbols and Abbreviations

TRADE SPECIFIC SYMBOLS

Lighting Outlets

○   Ceiling Surface Incandescent Fixture

⊢○   Wall Surface Incandescent Fixture

Ⓡ   Ceiling Recess Incandescent Fixture

⊢Ⓡ   Wall Recess Incandescent Fixture

$^A○_{3b}$ Standard Designation for All Lighting Fixtures – A = Fixture Type, 3 = Circuit Number, b = Switch Control

Ⓑ   Ceiling Blanked Outlet

⊢Ⓑ   Wall Blanked Outlet

Ⓔ   Ceiling Electrical Outlet

⊢Ⓔ   Wall Electrical Outlet

Ⓙ   Ceiling Junction Box

⊢Ⓙ   Wall Junction Box

Ⓛ$_{PS}$ Ceiling Lamp Holder with Pull Switch

⊢Ⓛ$_{PS}$ Wall Lamp Holder with Pull Switch

Ⓛ   Ceiling Outlet Controlled by Low Voltage Switching when Relay is Installed in Outlet Box

⊢Ⓛ   Wall Outlet – Same as Above

◇   Outlet Box with Extension Ring

EX→   Exit Sign with Arrow as Indicated

▢○   Surface Fluorescent Fixture

▢○▢P   Pendant Fluorescent Fixture

▢OR▢   Recessed Fluorescent Fixture

▢○▢   Wall Surface Fluorescent Fixture

⊟   Channel Mounted Fluorescent Fixture

▢○▢▢   Surface or Pendant Continuous Row Fluorescent Fixtures

▢OR▢▢   Recessed Continuous Row Fluorescent Fixtures

●   Incandescent Fixture on Emergency Circuit

▢•▢   Fluorescent Fixture on Emergency Circuit

Receptacle Outlets

⊢⊖   Single Receptacle Outlet

⊢⊖   Duplex Receptacle Outlet

⊢⊖ˣ   Duplex Receptacle Outlet "X" Indicates above Counter Max. Height = 42" or above Counter

⊢⊖$_{WP}$ Weatherproof Receptacle Outlet

⊢⊕   Triplex Receptacle Outlet

⊢⊕   Quadruplex Receptacle Outlet

⊢⊖   Duplex Receptacle Outlet – Split Wired

Receptacle Outlets (Cont).

Triplex Receptacle Outlet – Split Wired

Single Special Purpose Receptacle Outlet

Duplex Special Purpose Receptacle Outlet

Range Outlet

Special Purpose Connection – DW Dishwasher

Explosion-proof Receptacle Outlet Max. Height = 36" to ℄

Multi-outlet Assembly

Clock Hanger Receptacle

Fan Hanger Receptacle

Floor Single Receptacle Outlet

Floor Duplex Receptacle Outlet

Floor Special Purpose Outlet

Floor Telephone Outlet – Public

Floor Telephone Outlet – Private

Underfloor Duct and Junction Box for Triple, Double, or Single Duct System as Indicated by Number of Parallel Lines

Cellular Floor Header Duct

Switch Outlets

S Single Pole Switch, Max. Height = 42" to ℄

S_2 Double Pole Switch

S_3 Three-way Switch

S_4 Four-way Switch

S_D Automatic Door Switch

S_K Key Operated Switch

S_P Switch & Pilot Lamp

S_{CB} Circuit Breaker

S_{WCB} Weatherproof Circuit Breaker

S_{MC} Momentary Contact Switch

S_{RC} Remote Control Switch (Receiver)

S_{WP} Weatherproof Switch

S_F Fused Switch

S_L Switch for Low Voltage Switching System

S_{LM} Master Switch for Low Voltage Switching System

S_T Time Switch

S_{TH} Thermal Rated Motor Switch

S_{DM} Incandescent Dimmer Switch

S_{FDM} Fluorescent Dimmer Switch

Switch & Single Receptacle

Switch & Double Receptacle

Special Outlet Circuits

Institutional, Commercial, & Industrial System Outlets

Nurses Call System Devices – any Type

Paging System Devices – any Type

Fire Alarm System Devices – any Type

Fire Alarm Manual Station – Max. Height = 48" to ℄

Fire Alarm Horn with Integral Warning Light

Fire Alarm Thermodetector, Fixed Temperature

Smoke Detector

Fire Alarm Thermodetector, Rate of Rise

Fire Alarm Master Box – Max. Height per Fire Department

Magnetic Door Holder

Fire Alarm Annunciator

Staff Register System - any Type

Electrical Clock System Devices – any Type

Public Telephone System Devices

Private Telephone System Devices

Watchman System Devices

Sound System, L = Speaker, V = Volume Control

Other Signal System Devices – CTV = Television Antenna, DP = Data Processing

Institutional, Commercial, & Industrial System Outlets (Cont).

Symbol	Description
[SC]	Signal Central Station
◢	Telephone Interconnection Box
(PE)	Pneumaticy Electric Switch
(EP)	Electric/Pneumatic Switch
GP	Operating Room Grounding Plate
P 6	Patient Ground Point – 6 = Number of Iacks

Panelboards

Symbol	Description
⊢▭⊣	Flush Mounted Panelboard & Cabinet
⊢▭	Surface Mounted Panelboard & Cabinet
▬	Lighting Panel
▨	Power Panel
▭	Heating Panel
⊠	Controller (Starter)
☐⌐	Externally Operated Disconnect Switch

Busducts & Wireways

Symbol	Description
[T] [T] [T]	Trolley Duct
[B] [B] [B]	Busway (Service, Feeder, or Plug-in)
[C] [C] [C]	Cable through Ladder or Channel
[W] [W] [W]	Wireway
[J]	Bus Duct Iunction Box

Electrical Distribution or Lighting System, Aerial, Lightning Protection

Symbol	Description
○	Pole
⟍○—	Street Light & Bracket
△	Transformer
——	Primary Circuit
- - - -	Secondary Circuit
— — —	Auxiliary System Circuits
—⟶	Down Guy
—•—	Head Guy
—○⟶	Sidewalk Guy
⊢—	Service Weather Head

Symbol	Description
⊕	Lightning Rod
—L—	Lightning Protection System Conductor

Residential Signaling System Outlets

Symbol	Description
▣	Push Botton
☐↘	Buzzer
⊲	Bell
⊲↘	Bell and Buzzer Combination
⟁	Annunciator
◀	Outside Telephone
◁	Interconnecting Telephone
▮◀	Telephone Switchboard
BT	Bell Ringing Transformer
D	Electric Door Opener
M	Maid's Signal Plug
R	Radio Antenna Outlet
CH	Chime
TV	Television Antenna Outlet
T	Thermostat

Underground Electrical Distribution or Lighting System

Symbol	Description
M	Manhole
H	Handhole
TM	Transformer - Manhole or Vault
TP	Transformer Pad
— - - -	Underground Direct Burial Cable
⊢⟶- -	Underground Duct Line
⟐	Street Light Standard Fed from Underground Circuit

CS	Carbon Steel	DX	Deluxe White; Direct Expansion	FBGS	Fiberglass
Csc	Cosecant			F. C., f. c.	Footcandles; Compressive Stress in Concrete; Extreme Compressive Stress
C.S.F.	Hundred Square Feet	dyn	Dyne		
CSI	Construction Specifications Institute	e	Eccentricity		
		E	Equipment Only; East		
CT	Current Transformer	Ea.	Each	f. c. c.	Face-centered cubic
CTS	Copper Tube Size	Econ.	Economy	FCC	Flat Conductor Cable
Cu	Cubic	EDP	Electronic Data Processing	F.E.	Front End
CU	Copper			FEP	Fluorinated Ethylene Propylene (Teflon)
Cu. Ft.	Cubic Foot	EDR	Equivalent Direct Radiation		
cw	Continuous Wave			F.G.	Flat Grain
C.W.	Cool White	Eq.	Equation	FHA	Federal Housing Administration
Cwt.	100 Pounds	Elec.	Electrician; Electrical		
C.W.X.	Cool White Deluxe	Elev.	Elevator; Elevating; Elevator Constructor	Fig.	Figure
C.Y.	Cubic Yard (27 cubic feet)			Fin.	Finish; Finished
C.Y./Hr.	Cubic Yards per Hour	EMT	Electrical Metallic Conduit; Thin Wall Conduit	Fixt.	Fixture
Cyl.	Cylinder			Fl. Oz.	Fluid Ounce
d	Penny (nail size)			Flr.	Floor
D	Deep; Depth; Discharge; Diameter	Eng.	Engine	F.M.	Frequency Modulation; Factory Mutual
		EPDM	Ethylene Propylene Diene Monomer		
Dis.; Disch.	Discharge			Fmg.	Framing
Db	Decibel	Eqhv.	Equipment Operator, Heavy	Fndtn.	Foundation
Dbl.	Double			FOB	Free on Board
DC	Direct Current	Eqlt.	Equipment Operator, Light	Fori.	Foreman, inside
Demob.	Demobilization			Foro.	Foreman, outside
d.f.u.	Drainage Fixture Units	Eqmd.	Equipment Operator, Medium	Fount.	Fountain
D.H.	Double Hung			fpm	Feet per Minute
DHW	Domestic Hot Water	Eqmm.	Equipment Operator, Master Mechanic	FPT	Female Pipe Thread
Diag.	Diagonal			Fr	Frame
Diam.	Diameter	Eqol.	Equipment Operator, Oilers	F.R.	Fire Rating
Dist.	Distance			FRK	Foil Reinforced Kraft
Distri.	Distribution	Equip.	Equipment	FRP	Fiberglass Reinforced Plastic
Dk.	Deck	ERW	Electric Resistance Welded		
D.L.	Dead Load; Diesel			FS	Forged Steel
Do.	Ditto	Est.	Estimated	FSC	Cast Body; Cast Switch Box
Dp.	Depth	esu	Electrostatic Unit		
DPST	Double Pole, Single Throw	EW	Each Way	Ft.	Foot; Feet
		EWT	Entering Water Temperature	Ftng.	Fitting
Dr.	Driver			Ftg.	Footing
Drink.	Drinking	Excav.	Excavation	Ft. Lb.	Foot Pound
DS	Double Strength	Exp.	Expansion	Furn.	Furniture; Furnish; Furnished
DSA	Double Strength A Grade	Ext.	Exterior		
DSB	Double Strength B Grade	Extru	Extrusion	FVNR	Full Voltage Non-Reversing
Dty.	Duty	f.	Fiber stress		
		F	Fahrenheit; Female; Fill	FVR	Full Voltage Reversing
DWV	Drain Waste Vent	Fab.	Fabricated	FXM	Female by Male

Fy.	Minimum Yield Stress of Steel	Hyd.; Hydr.	Hydraulic	kWh	Kilowatt-hour
g	Gram	Hz	Hertz (cycles)	L	Labor Only; Length; Long
G	Gauss	I.	Moment of Inertia	Lab.	Labor
Ga.	Gauge	IC	Interrupting Capacity	lat	Latitude
Gal.	Gallon	ID	Inside Diameter; Inside Dimension; Identification	Lath.	Lather
Gal./Min.	Gallon per Minute			Lav.	Lavatory
Galv	Galvanized	IF	Inside Frosted	Lb; #	Pound
Gen.	General	IMC	Intermediate Metallic Conduit	LB	Load Bearing; L Conduit Body
Glaz.	Glazier	In.	Inch	L. & E.	Labor and Equipment
GPD	Gallons per Day	Incan.	Incandescent	lb./hr.	Pounds per Hour
GPH	Gallons per Hour	Incl.	Included; Including	lb./L.F.	Pounds per Linear Foot
GPM	Gallons per Minute	Ins.	Insurance	lbf/sq.in.	Pound-force per Square Inch
GR	Grade	Int.	Interior		
Gran	Granular	Inst.	Installation	LCL	Less than Carload Lot
Grnd	Ground	Insul.	Insulation; Insulate	Ld	Load
H	High; High Strength Bar Joist; Henry	Inter.	Interior	L.F.	Linear Foot
		IP	Iron Pipe	Lg.	Long; Length; Large
H.C.	High Capacity	IPS	Iron Pipe Size	L. & H.	Light and Heat
H.D.	Heavy Duty; High Density	IPT	Iron Pipe Threaded	L.H.	Long Span High Strength Bar Joist; Labor Hour
		J	Joule		
H.D.O.	High Density Overlaid	J.I.C.	Joint Industrial Council	L.J.	Long Span Standard Strength Bar Joist
Hdr.	Header	K	Thousand; Thousand Pounds		
Hdwe., Hdwr.	Hardware			LL	Live Load
		KAH	Thousand Ampere Hours	LLD	Lamp Lumen Depreciation
Help.	Helper average				
HEPA	High Efficiency Particulate Air	kemil	Thousand Circular Mils	lm	Lumen
		KDAT	Kiln Dried After Treatment	lm/sf	Lumen per Square Foot
Hg	Mercury			lm/W	Lumen per Watt
HID	High-Intensity Discharge	kg	Kilogram	L.O.A.	Length Overall
HO	High Output	k	Kilogauss	log	Logarithm
Horiz.	Horizontal	kgf	Kilogram Force	LP	Liquefied Petroleum; Low Pressure
HP	Horsepower; High Pressure	kHz	Kilohertz		
		Kip	1000 pounds	LPF	Low Power Factor
H.P.F.	High Power Factor	kj	Kilojoule	LPS	Low-pressure Sodium
Hr.	Hour	K.L.	Effective Length Factor	Lt.	Light
Hrs./Day	Hours per Day	KLF	Kips per Linear Foot	Lt. Ga.	Light Gauge
HSC	High Short Circuit	km	Kilometer	LTL	Less than Truckload Lot
Ht.	Height	KO	Knockout	Lt. Wt.	Lightweight
Htg.	Heating	KSF	Kips per Square Foot	LV	Low Voltage
Htrs.	Heaters	KSI	Kips per Square Inch	m	Meter
HVAC	Heating, Ventilating, and Air Conditioning	kV	Kilovolt	M	Thousand; Material; Male; Medium Wall Copper
		kVA	Kilovolt Ampere		
Hvy.	Heavy	kVAR	Kilovar (Reactance)	mA	Milliampere
HW	Hot Water	kW	Kilowatt		

Mach.	Machine	MRT	Mile Round Trip	ns	Nanosecond
Mag. Str.	Magnetic Starter	ms	Millisecond	NTS	Not to Scale
Maint.	Maintenance	M.S.F.	Thousand Square Feet	nW	Nanowatt
Marb.	Marble Setter	Mstz.	Mosaic and Terrazzo Worker	OB	Opposing Blade
Mat.	Material			O.C.	On Center
Mat'l.	Material	M.S.Y.	Thousand Square Yards	OD	Outside Diameter
Max.	Maximum	Mtd.	Mounted	O.D.	Outside Dimension
MBF	Thousand Board Feet	Mthe.	Mosaic and Terrazzo Helper	ODS	Overhead Distribution System
MBH	Thousand Btu's per Hour				
MC	Metal Clad	Mtng.	Mounting	O & P	Overhead and Profit
MCC	Motor Control Center	Mult.	Multiple; Multiply	Oper.	Operator
M.C.F.	Thousand Cubic Feet	MV	Megavolt; Mercury Vapor	Opng.	Opening
M.C.F.M.	Thousand Cubic Feet per Minute	MVA	Million Volt Amperes	Orna.	Ornamental
		MVAR	Million Volt Amperes Reactance	OS&Y	Outside Screw and Yoke
MCM	Thousand Circular Mils			Ovhd.	Overhead
MCP	Motor Circuit Protector	MW	Megawatt	Oz.	Ounce
MD	Medium Duty	MXM	Male by Male	P	Pole; Applied Load; Projection
M.D.O.	Medium Density Overlaid	MYD	Thousand Yards		
Med.	Medium	N	Natural; North	p.	Page
MF	Thousand Feet	nA	Nanoampere	Pape.	Paperhanger
M.F.B.M.	Thousand Feet Board Measure	NA	Not Available; Not Applicable	PAR	Weatherproof Reflector
				Pc.	Piece
Mfg.	Manufacturing	NBC	National Building Code	P.C.	Portland Cement; Power Connector
Mfrs.	Manufacturers	N.C.	Normally Closed		
mg	Milligram	N.E.C.	National Electrical Code	P.C.F.	Pounds per Cubic Foot
MGD	Million Gallons per Day	NEMA	National Electrical Manufacturers Association	P.E.	Professional Engineer; Porcelain Enamel; Polyethylene; Plain End
MGPH	Thousand Gallons per Hour				
MH	Manhole; Metal Halide, Man Hour	NEHB	Bolted Circuit Breaker to 240V	Perf.	Perforate; Perforated
		N. L. B.	Non-Load-Bearing	P.F.	Power Factor
MHz	Megahertz	nm	Nanometer	Ph.	Phase
Mi.	Mile	NM	Nonmetallic Cable	P.I.	Pressure Injected
M.I.	Malleable Iron; Mineral Insulated	NMC	Nonmetallic Cable with Corrosion-Resistant Sheath	Pile.	Pile Driver
				Pkg.	Package
mm	Millimeter			Pl.	Plate
Mill.	Millwright	No.	Number	Plah.	Plasterer Helper
Min.	Minimum; Minute	N. O.	Normally Open	Plas.	Plasterer
Misc.	Miscellaneous	N.O.C.	Not Otherwise Classified	Pluh.	Plumber's Helper
ml	Milliliter	Nose.	Nosing	Plum.	Plumber
M.L.F.	Thousand Linear Feet	N.P.T.	National Pipe Thread	Ply.	Plywood
Mo.	Month	NQOB	Bolted Circuit Breaker to 240V	p.m.	Post Meridiem
Mobil.	Mobilization			Pord.	Painter, Ordinary
Mog.	Mogul Base	NRC	Noise Reduction Coefficient	pp.	Pages
MPH	Miles per Hour			PP; PPL	Polypropylene
MPT	Male Pipe Thread	N.R.S.	Non Rising Stem	PPM	Parts per Million

Pr.	Pair	R.O.W.	Right of Way	Sswk.	Structural Steel Worker
Prefab.	Prefabricated	RPM	Revolutions per Minute	Sswl.	Structural Steel Welder
Prefin.	Prefinished	R.R.	Direct Burial Feeder	St.; Stl.	Steel
Prop.	Propelled		Conduit	STC	Sound Transmission
PSF; psf	Pounds per Square Foot	R.S.	Rapid Start		Coefficient
PSI; psi	Pounds per Square Inch	RT	Round Trip	Std.	Standard
PSIG; psig	Pounds per Square Inch Gauge	S	Suction; Single Entrance; South	Ston.	Stone Mason
PSP	Plastic Sewer Pipe	Scaf.	Scaffold	STP	Standard Temperature and Pressure
Pspr.	Painter, Spray	Sch.; Sched.	Schedule	Stpi.	Steamfitter; Pipefitter
Psst.	Painter, Structural Steel	S.C.R.	Modular Brick	Str.	Strength; Starter; Straight
PT	Potential Transformer	S.D.R.	Standard Dimension Ratio	Strd.	Stranded
PTZ	Pan, Tilt, and Zoom	sdwk.	Sidewalk	Struct.	Structural
P&T	Pressure and Temperature	S.E.	Surfaced Edge	Sty.	Story
Ptd.	Painted	SE;	Service Entrance Cable	Sub.	Subcontractor
Ptns.	Partitions	S.E.R.;		Subj.	Subject
Pu	Ultimate Load	S.E.U.		Surf.	Surface
PVC	Polyvinyl Chloride	S.F.	Square Foot	Sw.	Switch
Pvmt.	Pavement	S.F.C.A.	Square Foot Contact Area	Swbd.	Switchboard
Pwr.	Power	S.F.G.	Square Foot of Ground	S.Y.	Square Yard
Q	Quantity Heat Flow	S.F. Hor.	Square Foot Horizontal	Syn.	Synthetic
Quan.; Qty.	Quantity	S.F.R.	Square Feet of Radiation	Sys.	System
Q. C.	Quick Coupling	S.F. Shlf.	Square Foot of Shelf	t.	Thickness
r	Radius of Gyration	S4S	Surface 4 Sides	T	Temperature; Ton
R	Resistance	Shee.	Sheet Metal Worker	Tan	Tangent
R.C.P.	Reinforced Concrete Pipe	Sin	Sine	T.C.	Terra Cotta
Rect.	Rectangle	Skwk.	Skilled Worker	T.D.	Temperature Difference
Reg.	Regular	SL	Saran Lined	TFE	Tetrafluorethylene (Teflon)
Reinf.	Reinforced	S.L.	Slimline		
Req'd.	Required	Sldr.	Solder	T. & G.	Tongue & Groove; Tar & Gravel
Resi	Residential	S.N.	Solid Neutral		
Rgh.	Rough	SNM	Shielded Type NM Cable	Th.; Thk.	Thick
R.H.W.	Rubber, Heat, & Water Resistant; Residential Hot Water	S.P.	Static Pressure; Single Pole; Self-Propelled	Thn.	Thin
				Thrded.	Threaded
		Spri.	Sprinkler Installer	Tilf.	Tile Layer, Floor
rms	Root Mean Square	Sq.	Square; 100 square feet	Tilh.	Tile Layer Helper
Rnd.	Round	S.P.D.T.	Single Pole, Double Throw	THW	Insulated Strand Wire
Rodm.	Rodman (reinforcing)	S.P.S.T.	Single Pole, Single Throw	THWN;	Nylon Jacketed Wire
Rofc.	Roofer, Composition	SPT	Standard Pipe Thread	THHN	
Rofp.	Roofer, Precast	Sq. Hd.	Square Head	T.L.	Truckload
Rohe.	Roofer Helpers (Composition)	SS	Single Strength; Stainless Steel	Tot.	Total
				T.S.	Trigger Start
				Tr.	Trade
Rots.	Roofer, Tile & Slate	SSB	Single Strength B Grade	Transf.	Transformer

Trhv.	Truck Driver, Heavy	VA	Volt/amp	WT, Wt.	Weight
Trlr.	Trailer	V.A.T.	Vinyl Asbestos Tile	WWF	Welded Wire Fabric
Trlt.	Truck Driver, Light	VAV	Variable Air Volume	XFMR	Transformer
T.V.	Television	Vent.	Ventilating	XHD	Extra Heavy Duty
T.W.	Thermoplastic Water	Vert.	Vertical	XHHW	Cross-linked
	Resistant Wire	V.G.	Vertical Grain		Polyethylene Wire
UCI	Uniform Construction	V.H.F.	Very High Frequency		Insulation
	Index	VHO	Very High Output	Y	Wye
UF	Underground Feeder	Vib.	Vibrating	yd.	Yard
U.H.F.	Ultra High Frequency	V.L.F.	Vertical Linear Foot	yr.	Year
U.L.	Underwriters Laboratory	Vol.	Volume	Δ	Delta
Unfin.	Unfinished	W	Wire; Watt; Wide; West	%	Percent
UPS	Uninterruptible Power	w/	With	~	Approximately
	Supply	W.C.	Water Column; Water	ø	Phase
URD	Underground Residential		Closet	@	At
	Distribution	W.F.	Wide Flange	#	Pound; Number
USE	Underground Service	W.G.	Water Gauge	<	Less than
	Entrance Cable	Wldg.	Welding	>	Greater Than
UTP	Unshielded Twisted Pair	Wrck.	Wrecker; Wrecking		
V	Volt	W.S.P.	Water, Steam, Petroleum		

Index